TABLETOP MATH GAMES COLLECTION

46 Ways to Play Math

Math Games for All Ages

Volume One

Denise Gaskins

Tabletop Academy Press

Contents

Learning Math Through Play 5
Mental Math Do's and Don'ts 6
Journaling With Games 7

The Best Math Game Ever 9
The Substitution Game 11
Exhaust the Relationships 15
For More Information 20

Early Counting Games 21
Collect Ten 23
Chopsticks 25
Up and Down the Stairs 27
Domino Match 29
Dinosaur Race 31
Number Train 33
Fill the Stairs 35
For More Information 36

Counting Games with Bigger Numbers 37
Twenty-One 39
The Calendar Game 41
Fifty-Sticker Race 43
The Nickel Game 47

Copyright © 2023 Denise Gaskins
All rights reserved.
Tabletop Academy Press, Boody, IL, USA, TabletopAcademyPress.com.

Educational Use: You have permission to copy the math games in this book for personal, family, or single-classroom use only. Please enjoy them with your homeschool students, math circle, co-op, or other small local group.

About the Font: The Cadman font was designed by P.J. Miller to be as reader-friendly as possible. Letters are distinct and easy to distinguish, especially those most often confused by children and adults with dyslexia.

Two-Digit Number Train 53

Snugglenumber 55

Horseshoes 59

For More Information 60

Creative Nim 61

What Is a Nim Game? 63

Launch-It Nim 65

Traditional (Misère) Nim 69

Nim Variations 71

Race to the Pharaoh's Treasure 73

Make Your Own Nim Game 77

Early Mental Math 81

Coin Chain 83

Make and Take 85

Tiguous 87

Tiguous-Tac-Toe 93

Hit Me 95

Bowling 97

For More Information 102

Intermediate Mental Math 103

Distributive Dice 105

Leapfrog 109

Averages 115

Twenty-Four 117

Twenty-Four Variations 119

Contig 121

For More Information 124

The Function Machine 125
What Is a Function? 127
The Function Machine Game 128
Teaching Tips 129
Function Cards 130

Advanced Mental Math 139
Masquerade 141
Fight for the Center 143
What Two Numbers? 145
Greater Than 147
Power Up 151
Exponent Number Train 155
For More Information 156

Graphing Games 157
Hidden Hexagons 159
Coordinate Gomoku 163
Linear War 167
Radar 171
Radian Race 177
Racetrack 181
For More Information 188

Number Neighborhoods 189
Which Number Where? 191
Number Neighborhoods 195
For More Information 201

Special Thanks 202

Playful Math Books by Denise Gaskins 203

Learning Math Through Play

Clear off a table, find some dice or a deck of cards, and you're ready to enjoy playing math. Keep these tips in mind:

- Game rules are a social convention, easy to change by agreement among the players. Feel free to invent your own rules, and encourage players to modify the games as they play. As they tinker with the game, it prompts them to think more deeply about the math concepts involved.

- Rating math games by grade level is inherently arbitrary. Young players may eagerly join a game with advanced concepts if the fun of the challenge outweighs the work involved. On the other hand, don't worry that a game is too easy for anyone—even adults—as long as they find it interesting. Everyone benefits from a little extra practice in math, but it's the logic of strategy that makes a game fun.

- If you are a parent, these games provide opportunities to enjoy quality time with your children. If you are a classroom teacher, use the games as warm-ups and learning center activities or for a relaxing review day at the end of a term. If you are a tutor or homeschooler, make games a regular feature in your lesson plans to build your students' mental math skills.

- Try to let people learn by playing. Explain the rules as simply as possible and jump into the fun. You can add details, exceptions, and special situations as they come up during play or before starting future games.

- Be warned: Although children can play most of these games on their own, they learn much more if adults play along. When adults join the game, they reinforce the value of mathematical play.

- Talk as you play, especially with kids. As you watch your children's responses and listen to their comments, you'll discover what they understand about math. Where do they get confused? What do they do when they're stuck? Can they use the number patterns they've mastered to figure out something they don't know? How easily do they give up?

    Real education, learning that sticks for a lifetime, comes through person-to-person interactions. Children absorb more from the give and take of discussion with an adult than from even the best workbook or teaching video.

Mental Math Do's and Don'ts

Math games stretch young players' abilities to manipulate numbers in their heads. But please don't treat these games as worksheets in disguise. A game should be voluntary and fun. No matter how good it sounds to you, if a game doesn't interest your family, put it away. You can always try another one tomorrow.

You'll know when you find the perfect game because your children will wear you out wanting to play it again and again and again.

As you play, remember these tips for mental calculation:

- Don't just count. Limit straight counting to a few steps, so you can't lose track. That means you may work 39 + 2 by counting, but not 39 + 7. When you do count, always start at the bigger number, so you have fewer steps.

- Do break numbers apart. Work with the easier parts first.

- Do use logic to rearrange your numbers and simplify your calculation. For instance, to find 39 + 7 you can imagine moving one piece from the seven to the big pile: 39 + 7 is the same as 40 + 6.

- Don't try to keep track of "borrowing" or "carrying" numbers while you work. But you can use funny numbers as an intermediate step. For example, if you remember that 9 + 7 = 16, then you might think of 39 + 7 as thirty-sixteen.

- It often helps to add or subtract more than you really need, then adjust the calculation to fit. So instead of trying to add 39 + 7, you can add 40 + 7 and then take away the extra one.

- Don't try to memorize everything. Do memorize a few basic facts (such as the pairs of numbers that make ten) that you can use to figure out other things.

- Don't use the same trick on every calculation. Be creative, looking for new ways to use number relationships as you figure things out.

- Do use fingers, manipulatives, or marks on paper to keep track of information while you work, especially with longer, multistep calculations.

- Do allow your children plenty of time to think. Don't worry if a child stares blankly into space. That's often what "thinking hard" looks like. Try not to break their concentration.

Journaling With Games

Games are the ultimate re-playable activity prompts. As you repeat a game, you can try variations on your previous moves to gain extra advantage. This mirrors the approach a mathematician may take when faced with a problem. What if we try this, or that? How do things change, and what stays the same?

After you master the ordinary version of a game, try a *misère* variation. In a *misère* game, the move that otherwise would win now makes you the loser. Players must reconsider their strategy and think more deeply about the game.

Consider other ways to modify the game rules. Write about your ideas:

- ♦ If the game uses playing cards, can you figure out a way to play it with dice or dominoes? Or transfer it to a gameboard?

- ♦ What if we changed the number of cards to draw, or how many dice to throw on each turn?

- ♦ Or is there a way to use money in the game? Or can you change it into a whole-body action game? Perhaps using sidewalk chalk?

Older players may want to analyze a game, which can make a great writing prompt. What do you notice about the game, and what does it make you wonder?

- ♦ Does one player have the advantage, or do both players have an equal chance of winning?

- ♦ What's the best move? Can you find a strategy to increase your odds?

- ♦ Are fairness and randomness linked? Why or why not?

Unschooling advocate Pam Sorooshian explains the connection between games and math this way:

"Mathematicians don't sit around doing the kind of math that you learned in school. What they do is 'play around' with number games, spatial puzzles, strategy, and logic. They don't just play the same old games, though. They change the rules a little, and then they look at how the game changes.

"So, when you play games, you are doing exactly what mathematicians really do—if you fool with the games a bit, experiment, see how the play changes if you change a rule here and there. Oh, and when you make up games and they flop, be sure to examine why they flop—that is a big huge part of what mathematicians do, too."

All of the games in the *Tabletop Math Games Collection* include a matching page for writing your notes and modifications.

The Best Math Game Ever

CREATIVE MATH FOR ALL AGES

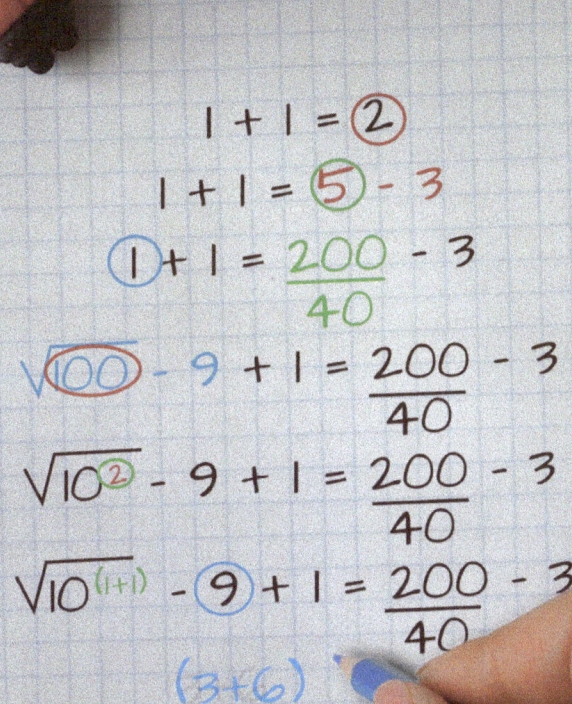

$1 + 1 = 2$

$1 + 1 = 5 - 3$

$1 + 1 = \dfrac{200}{40} - 3$

$\sqrt{100} - 9 + 1 = \dfrac{200}{40} - 3$

$\sqrt{10^2} - 9 + 1 = \dfrac{200}{40} - 3$

$\sqrt{10^{(1+1)}} - 9 + 1 = \dfrac{200}{40} - 3$

$(3+6)$

Notes & House Rules

The Substitution Game

MATH CONCEPTS: addition, subtraction, multiplication, division, order of operations, integers, fractions, equivalence and substitution.
PLAYERS: any number (a cooperative game).
EQUIPMENT: whiteboard and markers, or pencil and paper to share. Calculator optional.

Level One

The first player writes a simple equation at the top of the paper, such as "1 + 1 = 2." Then all players take turns complexifying this equation.

On your turn, copy the equation to the next line, replacing one number with an equivalent expression. For instance, replace the number 2 with:

$$5 - 3$$

$$\text{or } 50 \div 25$$

$$\text{or } (1/3) \times 6$$

… or any other calculation that equals two.

Use parentheses or brackets as needed to make your expression perfectly clear. For example, I put parentheses around my fraction above so people can tell I didn't mean "1/(3 × 6)," which is definitely not a substitute for two.

If you have colored pencils or markers, circle the number you plan to substitute. Then write the substitution below in the same color. Finally, fill out the rest of the equation using plain pencil or a black marker.

The other players should check to make sure they agree with your math.

After you change part of the equation, it is no longer available for anyone else to use. If you substitute "5 – 3" for the 2, the other players cannot replace your creation with their own version of that number. But they can alter individual numbers within your creation. So the next player may decide to substitute for the number 5, writing a new expression in its place.

Continue until the paper is full, or until the equation looks satisfyingly complex, or until you run out of time. Save the paper (or copy the final equation from the whiteboard) for playing Level Three.

$$1 + 1 = \boxed{2}$$

$$1 + 1 = \boxed{5} - 3$$

$$\boxed{1} + 1 = \frac{200}{40} - 3$$

$$\sqrt{\boxed{100}} - 9 + 1 = \frac{200}{40} - 3$$

$$\sqrt{10^{\boxed{2}}} - 9 + 1 = \frac{200}{40} - 3$$

$$\sqrt{10^{(1+1)}} - 9 + 1 = \frac{200}{40} - 3$$

A few rounds of the Substitution Game.

Players circled the number they wanted to change and wrote their substitution in the same color.

Then they copied the unchanged parts in neutral brown.

The Substitution Game

Level Two

As players grow comfortable with the basic game, pose additional challenges. Perhaps each substitution must use multiplication, or a fraction, or contain a specific number. If you're playing with a mixed-ability group, each player may have a different challenge.

For example, if your challenge is to use division, you might replace the number 3 in an existing equation with:

$$(70 \div 7) - 7$$

$$\text{or } (1/4) \times 4 \times 9 \div 3$$

... or any other calculation that includes division.

Use parentheses or brackets as needed to make your expression perfectly clear. The other players should check to make sure they agree with your math.

Continue until the paper is full, or until the equation looks satisfyingly complex, or until you run out of time. Save the paper (or copy the final equation from the whiteboard) for playing Level Three.

Level Three

The first player chooses an old round of the Substitution Game and writes the final equation at the top of the paper or whiteboard.

Players take turns simplifying the equation.

On your turn, copy the equation to the next line, replacing part of it with a simpler expression. The other players should check to make sure they agree with your math.

Continue until you reach the simplest form of the equation—which may not be the same as what the original game started with. For example, the simplest form of the equation "1 + 1 = 2" would be "2 = 2."

If the final equation is a true statement, then you win. Hooray!

But if anyone made a mistake in either level of play—either complexifying or simplifying the equation—you may end up with a nonsense statement like "2 = 13." Don't worry about trying to find the error. All the players still did plenty of mathematical thinking, so count this as a sideways win. Accept the reminder that you are human, and enjoy a good laugh at your silly result.

Notes & House Rules

Exhaust the Relationships

(The Substitution Game, Algebra-Style)

MATH CONCEPTS: addition, subtraction, multiplication, division, fractions, variables, equivalence and substitution.
PLAYERS: any number (a cooperative game).
EQUIPMENT: pencils and paper, or whiteboard and dry-erase markers. Cuisenaire rods, Legos, or other math blocks (or graph paper for drawing rectangles to scale).

Set-Up

Choose blocks of different colors and lengths to represent algebraic variables. Try to find blocks that have a relatively simple length relationship, such as "one block is twice as long as the other." If you have more than one block the same color, they should also have the same length.

Arrange two or three same-length rows of blocks. This models an algebraic equation because every row is equal to each of the other rows. For example, with Cuisenaire rods, we might build a pattern with the blue, light green, and dark green rods. (If you don't have blocks to arrange, you can draw a similar diagram with colored rectangles.)

Give each size of block a variable name, such as G for light green and D for dark green. Variable names may be capital letters or lower case, your choice. On your paper or whiteboard, make a chart with one column for each size, labeled with the appropriate variable name.

How To Play

Take turns noticing and naming relationships between your blocks. Write an expression for each relationship on your chart.

In my picture above, there are only three basic relationships:

- ♦ Two light green rods are the same length as one dark green.
- ♦ Three light green rods are the length of a blue rod.
- ♦ The difference between the blue and the dark green rods is one light green rod.

But as you will see, there are nearly infinite ways to describe these relationships using mathematical symbols.

Name the Relationships

Begin by writing the simplest patterns. For example, my dark green rod is twice as long as the light green one. So in the D column, I might list two expressions, two ways of stating that relationship:

G + G
2 × G [or simply 2G]

Look at the relationship from the other direction: The light green rod is half as long as the dark green one. And the difference between dark and light green is the length of one light green. So under the G label, I can list:

½ × D [or simply ½ D]
D − G

Or looking at the blue rod, I may write these combinations in the B column:

G + D
3G

Can you see how I could describe these same relationships from the perspective of the green or dark green rods?

D = B − G
G = B − D
G = ⅓ × B [or = ⅓ B]

These are the basic relationships between my three rods.

G	D	B
1/2 × D	G + G	G + D
D − G	2 G	3 × G
B − D	B − G	
1/3 × B		

Complexify the Relationships

When you have exhausted the basic links between your blocks, it's time to play. Algebra is all about playing with the idea of equivalent relationships. So dig in and have fun.

You can keep adding new relationships to the columns of your chart. But as the expressions grow more complex, they take up a lot of space. I like to write my fanciest creations below the chart (or on a new sheet of paper) as equations.

For example, I might combine the following relationships:

$$D = G + G$$
$$D = 2G$$
$$G = \tfrac{1}{2} D$$

... to create any of these equations:

$$D = \tfrac{1}{2} D + \tfrac{1}{2} D$$

$$G = \tfrac{1}{2} (G + G)$$

$$D = \tfrac{1}{2} (G + G) + \tfrac{1}{2} (2G)$$

$$G = \tfrac{1}{2} [G + \tfrac{1}{2} (2G)]$$

$$G = \tfrac{1}{2} \times 2 \times \tfrac{1}{2} \times 2 \times \tfrac{1}{2} \times 2G$$

etc.

As players take turns writing expressions and equations, the others check for mistakes. If you're not sure, compare the algebra to the original blocks. Don't accept any math until you convince yourself it's true.

Build wilder and sillier algebraic monstrosities for as long as everyone wants to play, or until you run out of time. But keep in mind that short lessons work best in math. Thinking hard can be just as tiring as a physical workout.

You can always come back to your chart and add more ideas tomorrow. Will you ever fully exhaust the relationships?

Variation

You can also play this as a solitaire game, just for the fun of complexifying equations. How crazy can you make the math?

Try It Yourself

Can you exhaust the relationships? Fill these pages with your expressions.

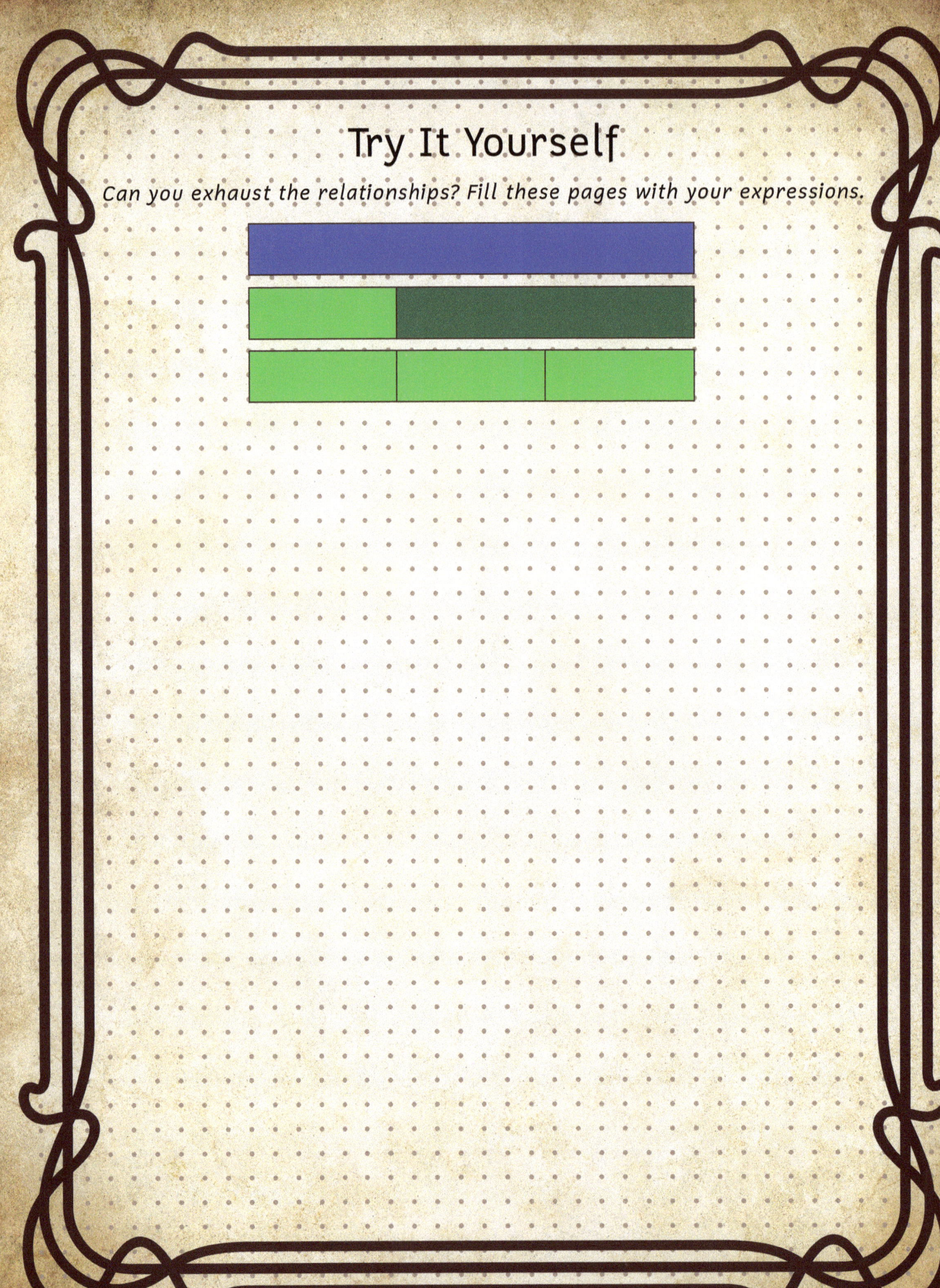

For More Information

Homeschooler Sonya Post calls herself a former "math witch" for how she used to teach math. Then she discovered the work of Caleb Gattegno and learned how to focus on the big ideas of math by helping children work with the key structures of algebra before they get lost in the details of arithmetic.

"I began my homeschooling journey in 1999 with my oldest," Post writes. "Oh, what I didn't know then! Math was a dreaded subject in our home. We shed lots of tears and lived under the tyranny of the textbook. Thankfully, we get to make up for past mistakes, and our children forgive us. I started over in 2015 with my youngest. This time, math is our favorite subject."

Now, through her Learn Math with Base Ten Blocks group on Facebook and the Learning Well At Home courses, Post equips parents to make sense of math so they can learn to teach their children.

- arithmophobianomore.com (to read about Post's journey of relearning math)
- facebook.com/groups/arithmophobianomore
- learningwellathome.com/build-and-learn-curriculum

Learn more about the Substitution Game and Exhaust the Relationships:

- arithmophobianomore.com/substitution-game-forget-worksheets
- arithmophobianomore.com/exhausting-relationships

◆ ◆ ◆

"The Substitution Game builds math from the ground up. The student is learning that there's nearly an infinite number of ways to write any number. Students are learning the 'secrets' to building complex mathematical statements, constantly relating new information to old understanding and taking one area of math and relating to another area.

"I did a training recently, and all we did is play the Substitution Game. Parents who were terrified of math suddenly began to see how this all works. They easily saw the relationships between the different areas of math.

"What hit me is that much of the math our students are learning in elementary and high school is just some variation of the substitution game. This game is worth more than 1,000 worksheets. A child hasn't mastered math when she completes those worksheets.

"You know what math a child has mastered when she can generate the math herself."

—Sonya Post, "Substitution Game: Forget the Worksheets"

TABLETOP MATH GAMES COLLECTION

EARLY COUNTING GAMES

7 WAYS TO PLAY MATH WITH YOUNG LEARNERS

Notes & House Rules

Collect Ten

MATH CONCEPTS: subitizing (at-a-glance counting), counting to ten.
PLAYERS: any number.
EQUIPMENT: counting cards (use the numbers 1–3 from the Tiny Polka Dot deck or cut corners off the lowest numbers from a deck of playing cards), a pile of small toys or other tokens, a bowl or basket for each player.

How To Play

Give each player a bowl or small basket. Place the pile of tokens (pennies, milk jug lids, cotton balls, or small toys) in the center, where all can reach. Spread the cards face down into a fishing pond.

On your turn, draw a card and count that many tokens to put in your bowl. Very young children can count by matching tokens to their card, placing one on each symbol. Now count how many items you have collected so far.

The first player to collect ten or more items is the winner. Or set your own counting target number.

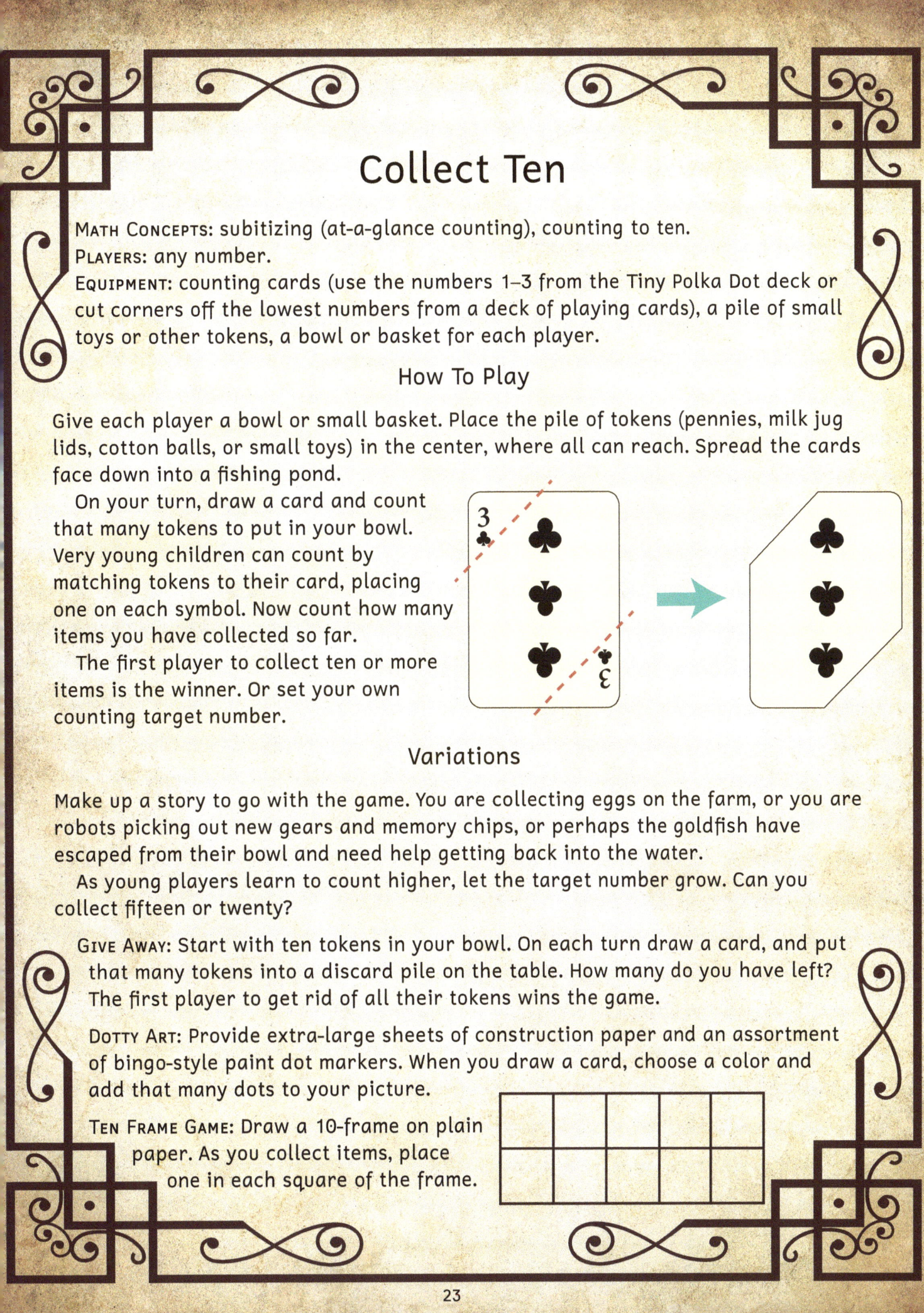

Variations

Make up a story to go with the game. You are collecting eggs on the farm, or you are robots picking out new gears and memory chips, or perhaps the goldfish have escaped from their bowl and need help getting back into the water.

As young players learn to count higher, let the target number grow. Can you collect fifteen or twenty?

GIVE AWAY: Start with ten tokens in your bowl. On each turn draw a card, and put that many tokens into a discard pile on the table. How many do you have left? The first player to get rid of all their tokens wins the game.

DOTTY ART: Provide extra-large sheets of construction paper and an assortment of bingo-style paint dot markers. When you draw a card, choose a color and add that many dots to your picture.

TEN FRAME GAME: Draw a 10-frame on plain paper. As you collect items, place one in each square of the frame.

Notes & House Rules

Chopsticks

Math Concepts: counting up to five, thinking ahead.
Players: two or more.
Equipment: none.

How To Play

Each player starts with both hands as fists, palm down, pointer fingers extended to show one point for each hand. On your turn, use one of your fingers to tap one hand:

- If you tap an opponent's hand, that person must extend as many extra fingers on that hand (in addition to the points already there) as you have showing on the hand that tapped. Your own fingers don't change.

- If you force your opponent to extend all the fingers and thumb on one hand (five or more points), that makes a *dead hand.* Dead hands go behind the player's back, out of the game.

- If you tap your own hand, you can *split* fingers from one hand to the other. For instance, if you have three points on one hand and only one on the other, you may tap hands to rearrange them, putting out two fingers on each hand. Splits do not have to end up even, but each hand must end up with at least one point (and less than five, of course).

- You may even revive a dead hand if you have enough fingers on your other hand to split. A dead hand has lost all its points, so it starts at zero. When you tap it, you can share out the points from your other hand as you wish.

The last player with a live hand wins the game.

Variations

House Rule: Do you want a shorter game? Omit the splits. Or you could allow ordinary splits but not splitting fingers to dead hands.

Nubs: All splits must share the fingers evenly between the hands. If you have an odd number of points, this will leave you with *half fingers,* shown by curling those fingers down.

Zombies: (For advanced players.) If a hand is tapped with more fingers than are needed to put it out of the game, the leftover points come back from the dead. If you have four fingers out, and your opponent taps you with a two-finger hand, that would fill up your hand with one point left over. Close your fist, and then hold out just the zombie point. The only way to kill a hand is to give it exactly five points.

Notes & House Rules

Up and Down the Stairs

Math Concepts: counting to ten, vectors (movement with direction).
Players: any number that fits on your stairway.
Equipment: six-sided die, a stairway or sidewalk.

How To Play

Play on any set of stairs where you won't get in other people's way. Or draw a series of chalk squares on a sidewalk or paved driveway. Or scratch a row of squares in the sand.

Start at the halfway step or square.

- Roll your die. Go up that number of steps.
- Roll again. This time, go down that number of steps.
- Keep rolling the die, alternating movements up and down.

Will you ever escape the stairs?

Variation

Use playing cards instead of rolling a die. Remove any playing cards from your deck that have a number equal to or greater than half the number of stairsteps. Have an adult (or non-player) draw cards and read numbers for the children to move.

What happens if you move up for a black card and down for a red card?

♦ ♦ ♦

Halfway down the stairs
Is a stair
Where I sit.
There isn't any
Other stair
   Quite like
   It.
      I'm not at the
      bottom,
         I'm not at the
         top;
         So this is the
         stair
         Where
         I always
         Stop.

Halfway up the stairs
Isn't up
And isn't down.
It isn't in the nursery,
It isn't in the town.
And all sorts of
funny thoughts
Run round my head:
"It isn't really
Anywhere!
It's somewhere else
Instead!"

—A.A. Milne
When We Were Very Young

Notes & House Rules

Domino Match

MATH CONCEPTS: counting to six, subitizing with dot patterns, visual memory, matching sets.
PLAYERS: two or more.
EQUIPMENT: one set of double-six dominoes, two six-sided dice.

How To Play

Remove domino tiles with blanks and set all remaining tiles face up in the middle of the table.

On your turn, roll the two dice and claim the domino tile that matches the dot patterns on both dice, unless that tile has already been taken by another player.

With two players, the first to claim six tiles wins. For three players the goal is four tiles, and for larger groups it takes three tiles to win the game.

Variations

HOUSE RULE: How will you handle duplicate rolls? At our house, if your dice match a tile someone else has already claimed, you can steal that tile from the other player.

DOMINO COVER-UP: Each player needs a specific type or color of token. When another person has already taken the tile that matches both dice rolled, the player may put tokens on any two separate tiles, covering the matching numbers. On a future turn, the player may be able to claim one of these tiles by rolling the uncovered number.

If you roll a two and three, then look for the domino tile with those numbers.

Notes & House Rules

Dinosaur Race

MATH CONCEPTS: number symbols, counting beyond ten, number line.
PLAYERS: any number.
EQUIPMENT: counting cards (use the numbers 1–3 from the Tiny Polka Dot deck or cut corners off the lowest numbers from a deck of playing cards), number line racetrack, small plastic dinosaur or other toy for each player.

Set-Up

Draw a straight path on paper or a manila file folder, either horizontal or slanted uphill (so the larger numbers will be higher). Divide the racetrack into twelve to twenty spaces large enough for small toys to rest in. Or glue squares of colored construction paper in a long line on poster board. Number the spaces in order, beginning with one.

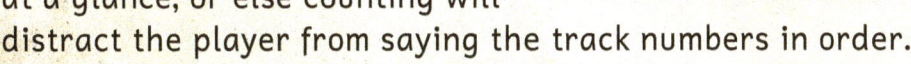

Use only the cards numbered 1–3. The number of squares moved each turn must be low enough to recognize at a glance, or else counting will distract the player from saying the track numbers in order.

How To Play

Each player should choose a small dinosaur or other toy and place it near the beginning of the racetrack. Turn the cards face down and spread them out to form a fishing pond.

On your turn, draw a card and move your dinosaur that many spaces, saying each number as you land on it. Cards should be mixed back into the pond after each turn. This is the most important rule: *when moving their toys, players must say the number in each space.* Repeating the numbers in order focuses attention and helps build number sense, a gut feeling for how numbers work, which is important to future learning.

The first player to reach the end of the path wins the race.

Variation

WHOLE-BODY COUNTING: Draw a Dinosaur Race path outdoors with sidewalk chalk, or use colored painter's tape along a hallway floor.

Notes & House Rules

Number Train

Math Concepts: number symbols, numerical order, thinking ahead.
Players: two or more.
Equipment: one deck of playing cards (face cards removed) or a double deck for more than four players, additional cards to use as train cars.

Set-Up

Give each player four to six miscellaneous cards to serve as the cars of their number trains. Lay these cards face down in a horizontal row, as shown.
Shuffle the deck of number cards and spread it on the table as a fishing pond.

House Rule: Decide how strict you will be about the "increases from left to right" rule and repeated numbers. Does "1, 3, 3, 7, 8" count as a valid number train? Or will the player have to keep trying for a card to replace one of the threes?

How To Play

On your turn, draw one card and play it face up on one of your train cars. The numbers on your train must increase from left to right, but they do not need to be in consecutive order. If you do not have an appropriate blank place for your card, you have two choices:

- Mix the new card back into the fishing pond.
- Use the new number to replace one of your other cards, and then mix the old one into the fishing pond.

The first player to complete a train of numbers that increases from left to right wins the game.

Line up the cars of your train.

Two of the train cars have passengers. Which numbers could you put on the other cars?

Notes & House Rules

Fill the Stairs

MATH CONCEPTS: counting to ten, negative numbers.
PLAYERS: any number.
EQUIPMENT: playing cards (face cards and jokers removed), pencil and paper.

How To Play

Each player draws 11 stairsteps (counting the top and bottom floors) on a piece of paper. Write zero on your middle step.

Agree on which color playing card represents negative numbers. Mix the cards face down in a fishing pond.

On your turn, choose one card. Write the number from your card on one of your stairsteps.

Then mix your card back into the pond.

The first player to fill their stairs with numbers in order wins the game.

The numbers have to grow as you go up the stairs and get smaller going down. But you can skip numbers. For instance, you could put +2 on the stair above zero, if you like. Or you could write –4 two steps below zero, leaving only one blank in between.

If you draw a card that will not fit on your stairs, you lose that turn. Mix the card back into the pond without writing anything.

If you make a mistake—like putting +9 too close to the middle, so there's no way to fill the higher steps—you can use a turn to erase one number on your stairs. You don't get to choose a new card on the same turn as erasing a number.

Comments

As soon as children are able to confidently count backwards, they can begin thinking about negative numbers. What happens when you keep counting down past zero?

Variations

For a faster game, draw two cards on each turn. Place their numbers on different stairsteps.

MENTAL MATH STAIRS: Draw two cards and write their sum (addition) or difference (subtraction) on one of your stairsteps. Players may choose whether they want their answer to be positive or negative, so the stair numbers can range from –18 to +18.

For More Information

"Halfway Down the Stairs," from *When We Were Very Young,* A. A. Milne, illustrated by E. H. Shepard. Methuen & Co, London, 1924.

Chopsticks

Finger-counting games are common in eastern Asia—and they must be contagious, since my daughters caught them from their Korean friends at college.

Middle school teacher Nico Rowinsky shared Chopsticks (which is simpler than the version my daughters brought home) in a comment on the "Tiny Math Games" post at Dan Meyer's blog.

♦ blog.mrmeyer.com/2013/tiny-math-games

Domino Match

Domino Cover-Up was created by Jean Carlton, lead teacher in the infant room at Old Dominion University Child Development Center, and shared by Alice P. Wakefield in *Early Childhood Number Games: Teachers Reinvent Math Instruction,* Allyn & Bacon, 1998.

Dinosaur Race

Counting up and down a number line forms a strong foundation for children's understanding of arithmetic. Dinosaur Race is based on the research of Robert S. Siegler and Geetha B. Ramani, who studied how preschool children responded to a variety of games.

Playing a number line game like Dinosaur Race for as little as an hour (in fifteen-minute segments spread out over a couple of weeks) made a dramatic difference in the children's ability to learn and retain arithmetic facts, while similar games played on a round track or on a linear track without numbers produced no measurable change.

Siegler, Robert S., and Geetha B. Ramani. "Playing Linear Number Board Games—But Not Circular Ones—Improves Low-Income Preschoolers' Numerical Understanding," Journal of Educational Psychology, vol. 101 (2009), no. 3, 545–560.

♦ psycnet.apa.org/record/2009-11043-002

TABLETOP MATH GAMES COLLECTION

COUNTING GAMES with BIGGER NUMBERS

7 WAYS TO PLAY MATH WITH PRIMARY STUDENTS

Notes & House Rules

Twenty-One

MATH CONCEPTS: counting to twenty-one, thinking ahead.
PLAYERS: two or more.
EQUIPMENT: none.

How To Play

The first player says "one" or "one, two" or "one, two, three."
  Each player then counts in turn, increasing the total by one, two, or three numbers. Whoever is forced to say twenty-one loses the game.
  If there are more than two players, the player who says twenty-one drops out of the game. The next person in the rotation starts a new round by counting from one, and play continues. The loser of each round drops out, until only one player (the winner) remains.

Variations

Choose a different poison number. Or allow a different number of counts per turn.
  Or count down from your chosen number, and the person who says zero loses the game.
  Or let the player who reaches the target number win the game.
  What other variations can you invent?

TWISTER: The first player rolls a six-sided die and says that number. Then each player in turn tips the die, turning a number that had been on the side up to the top, and then adds that number to the current total.
  The player who exactly reaches twenty-one, or who forces the next player to go over, wins the game.

Notes & House Rules

The Calendar Game

Math Concepts: counting to 31, months and days, thinking ahead.
Players: two or more.
Equipment: none.

How To Play

The first player says any date in January.

Then each player in turn increases either the month or the day (but never both at once) and says a new date later in the year.

Whoever says December 31 loses the game.

If you have more than two players, the player who says "December 31" drops out of the game. The next person in the rotation starts again with any date in January. Then the player left after everyone else has been eliminated wins the game.

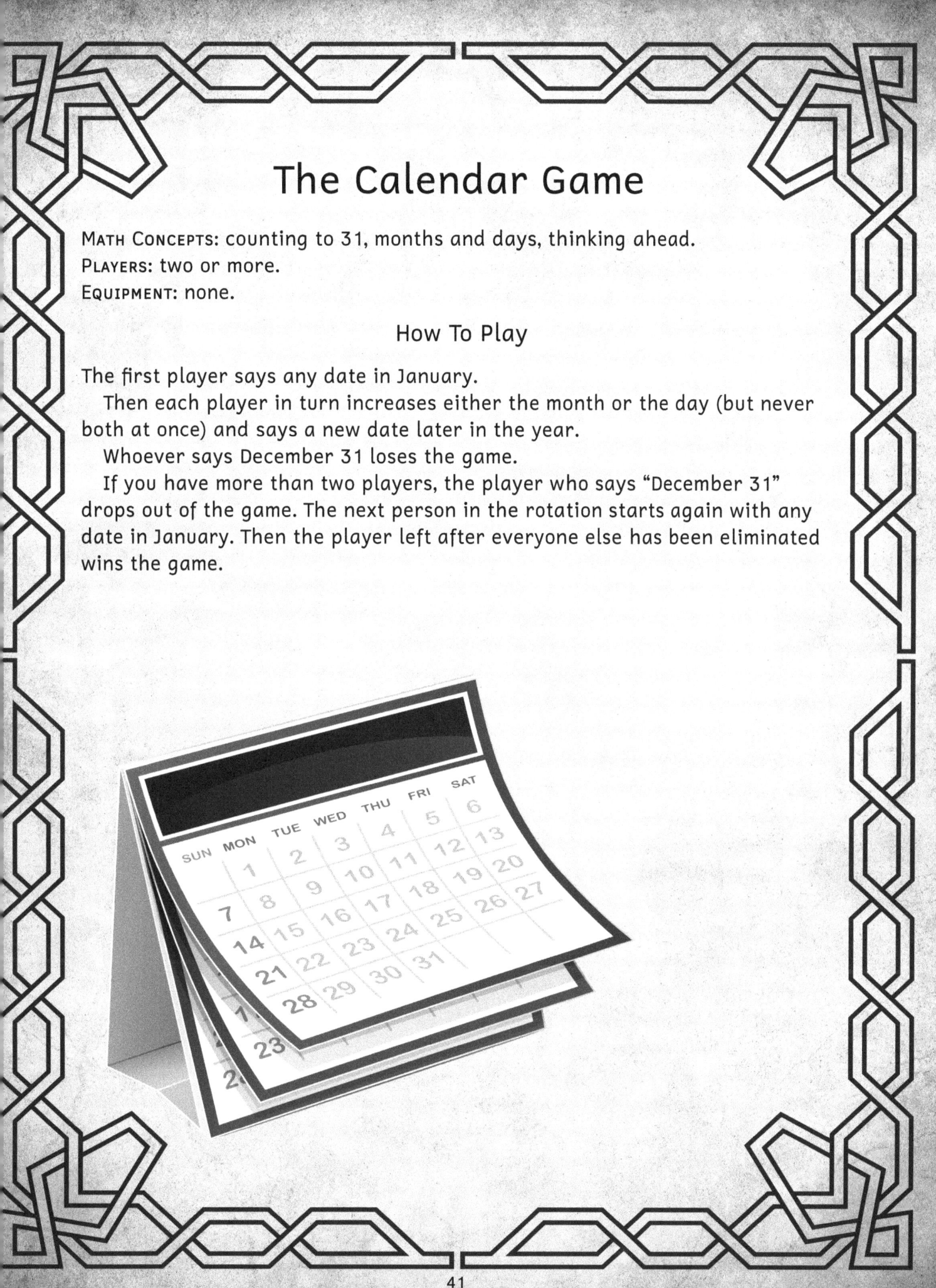

Notes & House Rules

Fifty-Sticker Race

MATH CONCEPTS: counting to fifty, how to read a hundred chart.
PLAYERS: any number.
EQUIPMENT: one gameboard for each player, one six-sided die, a sheet or roll of small stickers (fifty per player), scissors.

Set-Up

Each player will need a copy of the Fifty Chart gameboard and a roll or sheet of stickers. The fifty chart is like a number line cut up and laid in rows, top-down or bottom-up. Arrows guide the player from line to line.*

Players may each have a separate pair of scissors, or a shared pair may be passed around the table with the die.

By cutting the correct number of stickers from a roll or sheet, you separate the counting step from the number line move, so players can read the numbers out loud without the mental dissonance of having to count spaces.

How To Play

On your turn, roll the die and cut that many stickers from your sticker sheet.

Beginning with the number one, put a sticker on each square, saying aloud each number as you cover it. When you reach the end of a row, follow the arrow down to the next line of numbers.

The first player to cover all fifty squares wins the race. Other players may continue to take turns with the die until they fill their own charts.

*For free printable gameboards, download the Number Game Printables Pack at TabletopAcademy.net/free-printables.

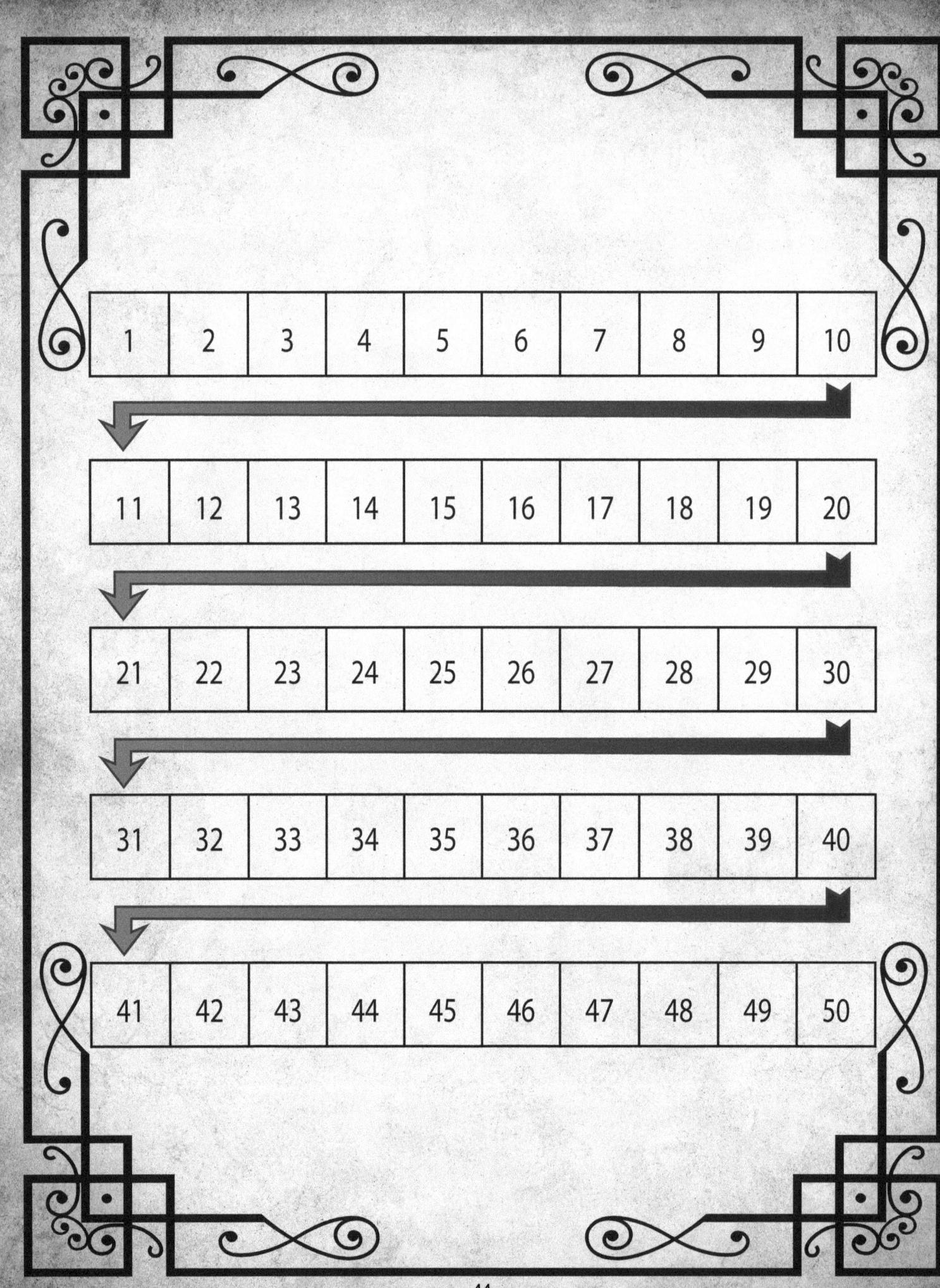

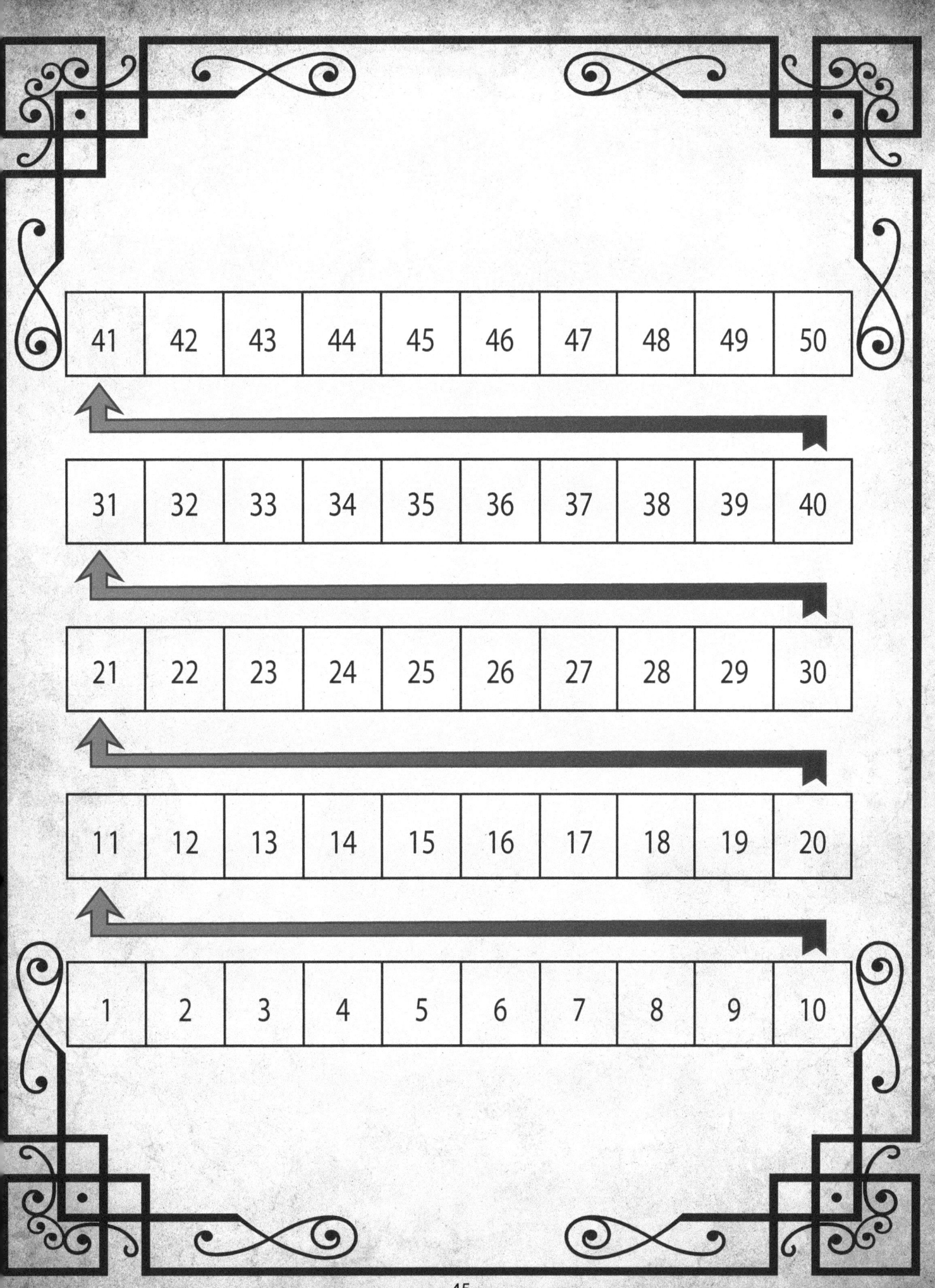

Notes & House Rules

The Nickel Game

MATH CONCEPTS: numbers to one hundred, greater than or less than, logical strategy.
PLAYERS: two players or two teams.
EQUIPMENT: hundred chart, twenty-four nickels or other small tokens.

Set-Up

Choose one chart for players to share. Many people find the bottoms-up hundred chart more logical than the traditional top-down version. It makes intuitive sense to have the numbers get larger as they climb up the page.

Each player starts with twelve nickels (5¢ coins) or other small tokens like dried beans or colored glass gems, or you can even use crumpled bits of paper.

How To Play

One player chooses a secret number from one to one hundred. The other player pays a nickel for each guess by placing it on a square of the hundred chart. The first player signals thumb-up if the secret number is greater than the guess or thumb-down if it is less.

Or you can practice math vocabulary. The first player can say, "My number is greater (or less) than [name the number guessed]."

The second player continues to buy guesses with nickels until the secret number is revealed. When the number is correct, the player who had the secret collects all the nickels from the chart.

Then the players trade roles. Play as many complete rounds as desired or until one player runs out of money.

Variation

Play orally for an easy travel game. If you want to keep score, count how many guesses it takes to find the other player's number.

This is a good place for finger counting, since you can hold up a finger for each guess without distracting your mind from the hunt for the secret number.

1	2	3	4	5	6	7	8	9	10
11	12	13	14	15	16	17	18	19	20
21	22	23	24	25	26	27	28	29	30
31	32	33	34	35	36	37	38	39	40
41	42	43	44	45	46	47	48	49	50
51	52	53	54	55	56	57	58	59	60
61	62	63	64	65	66	67	68	69	70
71	72	73	74	75	76	77	78	79	80
81	82	83	84	85	86	87	88	89	90
91	92	93	94	95	96	97	98	99	100

91	92	93	94	95	96	97	98	99	100
81	82	83	84	85	86	87	88	89	90
71	72	73	74	75	76	77	78	79	80
61	62	63	64	65	66	67	68	69	70
51	52	53	54	55	56	57	58	59	60
41	42	43	44	45	46	47	48	49	50
31	32	33	34	35	36	37	38	39	40
21	22	23	24	25	26	27	28	29	30
11	12	13	14	15	16	17	18	19	20
1	2	3	4	5	6	7	8	9	10

0	1	2	3	4	5	6	7	8	9
10	11	12	13	14	15	16	17	18	19
20	21	22	23	24	25	26	27	28	29
30	31	32	33	34	35	36	37	38	39
40	41	42	43	44	45	46	47	48	49
50	51	52	53	54	55	56	57	58	59
60	61	62	63	64	65	66	67	68	69
70	71	72	73	74	75	76	77	78	79
80	81	82	83	84	85	86	87	88	89
90	91	92	93	94	95	96	97	98	99

90	91	92	93	94	95	96	97	98	99
80	81	82	83	84	85	86	87	88	89
70	71	72	73	74	75	76	77	78	79
60	61	62	63	64	65	66	67	68	69
50	51	52	53	54	55	56	57	58	59
40	41	42	43	44	45	46	47	48	49
30	31	32	33	34	35	36	37	38	39
20	21	22	23	24	25	26	27	28	29
10	11	12	13	14	15	16	17	18	19
0	1	2	3	4	5	6	7	8	9

Notes & House Rules

Two-Digit Number Train

MATH CONCEPTS: place value, numerical order, thinking ahead.
PLAYERS: two or more.
EQUIPMENT: one deck of playing cards (tens, jacks, and kings removed) or two ten-sided dice, pencil and paper or whiteboard and markers.

Set-Up

Decide how long your number trains will be: five to ten spaces. Players draw their own number trains on paper or on a whiteboard.

A train may be any shape (a simple row of boxes, stairsteps, a caterpillar of ovals with legs, or a chain of flowers with open centers for writing in) and may curl around the page in any direction, but it must have a clear beginning and end. Each space must have enough room to write a two-digit number.

If you draw the trains on paper, you can laminate these drawings or slip them into sheet protectors for repeated play. But if you make a new drawing each time, then the Number Train game can grow longer and express your personality as your artistic skills develop.

How To Play

Shuffle well, and then spread the cards face down as a fishing pond.

On your turn, draw two cards (or roll the dice) and arrange them to make a two-digit number. You may use a zero (queen) as the tens digit, if you wish, which makes the equivalent of a one-digit number: Q6 = "06" = 6.

Write your number into any blank space in your train, making sure that the numbers increase from the beginning of your number train to the end, and then mix the cards back into the pond.

Be careful: once written, your number may not be erased.

If you cannot make a number that fits, discard. You have to wait until your next turn to try again. The first player to complete an ordered train of increasing numbers wins the game.

HOUSE RULE: Allow players to erase a number from their train. This option uses up the whole turn, so don't draw any cards.

Would you make the number 27, or is 72 a better choice?

Notes & House Rules

Snugglenumber

MATH CONCEPTS: place value, probability, thinking ahead.
PLAYERS: any number.
EQUIPMENT: one deck of playing cards (face cards removed) or a ten-sided die, pens or pencils, blank paper or gameboards.

Set-Up

You can print gameboards or write the numbers 0, 5, 10, 25, 50, 100 down the center of your paper. These are the *snugglenumbers* (target numbers).

Next to each number, draw as many blanks as there are digits in that snugglenumber. These blanks are where you do your snuggling. The gameboard has two columns of blanks, which can be used for two separate games. Or two players may share one page, each using one of the columns.

Remove the tens from your deck of cards and replace them with queens to represent the number zero—or leave in the tens, but count them as zeros. Shuffle the deck and place it in the center of the table where everyone can reach.

How To Play

On your turn, flip one card face up beside the deck (or roll the die). Each player must write that number on one of their blanks, trying to create numbers in each row that are as close to each snugglenumber as possible. The next player waits until everyone has filled a blank before turning up the next card.

Once you have written a digit, it cannot be moved. But the gameboard includes a trash can symbol, so once in each game you can decide to throw away a card, writing its number value in the can instead of on a blank.

When all the blanks are filled in, players compare their numbers. Whoever has the snuggliest number in each row gets a point. In the case of a tie—either the players made the same number, or they made two numbers that are equally close to the target—both players earn a point.

Whoever wins the most points wins the game.

Variation

For older students, the players subtract the numbers they made from the snugglenumbers—or vice versa, depending on which is bigger—and then add up all these differences. The player with the smallest total difference wins.

Snugglenumber: A Sample Game

Sven challenged Olaf to a game of Snugglenumber. Olaf drew first, turning up an ace. Both players wrote a one on their gameboard. Olaf put his by the zero. Since one is very close to zero, he thought he had a good chance of winning that row. Sven wrote his one in the hundreds place.

   Then Sven turned up a four, so both players found a place for that digit. Sven wrote his four next to the five—only one point away, a likely winner. Olaf put his four in the tens place next to the fifty, figuring a large number was bound to come along for the ones place.

   Olaf's turn to draw, and he got an eight. He wrote it next to the four, making forty-eight and snuggling very close to the fifty. Sven put the eight in the twenty-five row, hoping to draw a two later in the game.

   Sven turned up a six, and the players wrote it in. Then Olaf turned up a queen, which stands for a zero. Sven pounced on the chance to score on the zero row. Olaf put his zero in the hundreds place, hoping to draw nines later. Then Sven turned up another zero...

Olaf		Sven
_ 1	0	0 _
_ _	5	4 _
_ 0	10	_ _
_ 6	25	_ 8
4 8	50	_ _
0 _	100	1 0 6

___ 0 ___

___ 5 ___

___ 10 ___

___ ___ 25 ___ ___

___ ___ 50 ___ ___

___ ___ 100 ___ ___

Notes & House Rules

Horseshoes

(A Snugglenumber Variation)

Math Concepts: place value, strategic thinking.
Players: two or more.
Equipment: one deck of playing cards (face cards removed), pens or pencils and scratch paper for keeping score.

Set-Up

Separate out the cards numbered ace (one) through nine, plus cards to represent the digit zero. We use the queens (Q is round enough for pretend), but you could also use the tens and just count them as zeros.

How To Play

Shuffle well and deal eleven cards to each player.

Arrange your cards in the pattern shown here, one card in place of each blank line, to form numbers that come as close to each target number as you can get.

Score according to horseshoes rules:

- Three points for each ringer, or exact hit on the target.
- One point for each number that is six or less away from the target.
- If none of the players land in the scoring range for a target number, then score one point for the number closest to that target.

```
  0  __
  5  __
 10  __ __
 25  __ __
 50  __ __
100  __ __ __
```

For a quick game, whoever scores the most points wins.

Or follow horseshoes tradition and play additional rounds until one player gets 21 points (40 for championship games)—and you have to win by at least two points over your closest opponent's score.

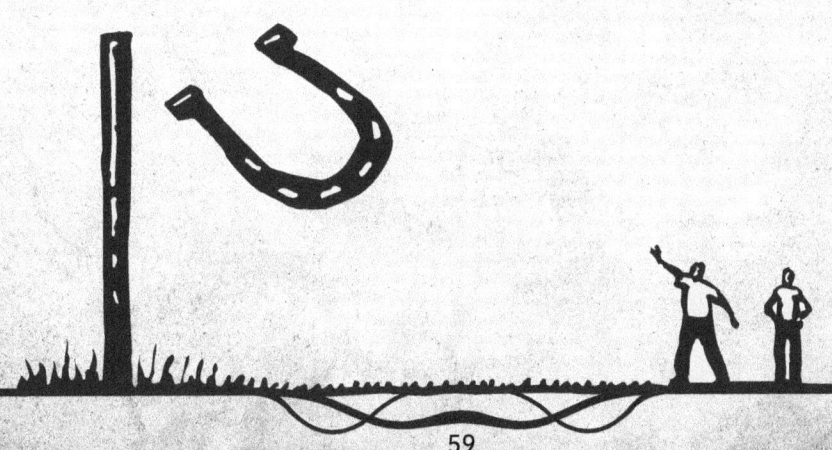

For More Information

Twenty-One and The Calendar Game

These games are variations of Nim, one of the oldest and most flexible math games in the world. Jim Pardun shared the Calendar Game in a comment on Dan Meyer's "Tiny Math Games" blog post.

♦ blog.mrmeyer.com/2013/tiny-math-games

Snugglenumber

I first saw place value games on the late-1980s PBS Square One Television series, which had a faux game show routine called "But Who's Counting?" Math teacher Anna Weltman posted this version at her blog Recipes for π.

"The game of Snugglenumber has taken my school by storm," Weltman writes. "Kids from third grade to tenth-grade Algebra 2 beg to play it. It involves the seemingly mundane arithmetic concept of place value. And yet, everyone loves it ... Oh, and did I mention that when you say Snugglenumber you must scrunch up your nose, smile adorably, and coo, 'Snug-gle-num-ber'?"

♦ recipesforpi.wordpress.com/2013/10/16/snugglenumber

"Math games are important because they reject an article of faith: that students must learn and practice the basic skills of mathematics before they can do anything interesting with them.

"People don't mind practicing a sport because playing the sport is fun. It's easy to tell a tennis player to practice 100 serves from the ad side of the court, for instance, because the tennis player has mentally connected the acts of practicing tennis and playing tennis.

"If you need to learn multiplication facts, one option is to watch a video and then drill away. Or we can queue up all that practice in a tiny math game that'll have students playing as they practice.

"Easy money says the student who's practicing math while playing it will practice more multiplication and enjoy that practice more than the student who is assigned to drill practice alone."

—Dan Meyer, "Tiny Math Games"

Creative Nim

MAKE YOUR OWN MATH & LOGIC GAMES

Notes & House Rules

What Is a Nim Game?

Nim is a pure strategy game for two players. On each turn, players remove an option until finally no choice remains.

Game options might include:

- How many stones to take from a pile.
- Which position to claim on a gameboard.
- How far to count in a given sequence.

The rules can vary at the players' whim (as long as both players agree). How many possibilities do you start with, what are the rules for removing options, and how do you win or lose the game? Everything is open to change. And with every tweak, players must reanalyze their strategy.

For centuries, people all around the world have played Nim-like folk games, though the rules are rarely written down. Some say Nim originated in China because the rules are similar to the Chinese game Jian-shízi, or "picking stones."

The first version of the game in print dates to about 1500 in a book of mathematical recreations by Luca Pacioli, a Franciscan friar who also collaborated with Leonardo da Vinci on a geometry book.

Charles Bouton coined the modern name (perhaps based on the German word for "to take") and brought mathematical attention to the game with his 1901 article "Nim, A Game with a Complete Mathematical Theory." But it was Martin Gardner who made the game famous when he wrote about Nim in his *Scientific American* column "Mathematical Games" in 1958.

How Is This Math?

Mathematicians enjoy studying patterns, whether they are patterns within our system of numbers or patterns of shapes or of abstract ideas.

Did you know there's a branch of mathematics called *game theory?* Game theorists study patterns of logical decision-making.

They began by studying the pattern of wins and losses in 2-player strategy games like Nim. Now the science has grown into a study of much larger and more complex "games" like politics and economics.

A Puzzle for You

Think about all the games you know. How many can you find that have the features of a Nim game?

Notes & House Rules

Launch-It Nim

MATH CONCEPTS: logic and strategic thinking.
PLAYERS: only two.
EQUIPMENT: gameboard, 10-12 toothpicks or other small tokens.

Traditionally, Nim is *a misère* (MEE-ZAIR) game. The player who takes the last stone loses the game. But with young children, I prefer making the last item the winner.

In my homeschool co-op math classes, we start by launching a few balloons on the whiteboard, the whole class against the teacher. I try to give new players at least two chances before I go for the win. Then the kids pair off to play against each other, taking turns on who goes first.

When students think they've figured out the balloon game, I challenge them to move to the rocket, which requires a slight modification in strategy. If they can't beat me on the rocket, I know they haven't really mastered the game, so I send them back to their tables for more practice.

Set-Up

For this game, you will need two players and a handful of toothpicks.

- To launch the hot air balloon, use 10 toothpicks.
- To launch the rocket, use 12 toothpicks.

If you don't have toothpicks, any set of small items will work. You can put a penny on each toothpick image, or use scraps of paper rolled to make stick-like shapes.

How To Play

Choose your gameboard. Children may want to color the picture before playing.

- Place toothpicks on the balloon's rope pegs or the rocket's launch tower as shown. Or play a purely abstract game with toothpicks in a single pile on the table.
- Allow the youngest player choice of moving first or second. In succeeding games, allow the loser of the previous game to choose.
- On your turn, remove one or two toothpicks from the picture. You must take at least one toothpick, but you may not take more than two.
- Whoever takes the last toothpick wins a ride on the balloon or the rocket.

Or play *a misère* game by making the last toothpick "poison." Whoever takes it loses the game. This adds a twist to the strategy, and it allows for an interesting endgame. Many students actually try to lose, just so they can act out a melodramatic death scene.

Launch the Hot Air Balloon

Notes & House Rules

Traditional (Misère) Nim

Math Concepts: logic and strategic thinking.
Players: only two.
Equipment: assorted game tokens, or pencil and paper.

Launch-It Nim is a favorite game at our math club meetings, but it is like Tic-Tac-Toe in that once you figure out the trick, you can almost always win (unless the other player knows, too). That makes for a boring game.

For a greater challenge, try the traditional multi-pile version of Nim.

Traditional Nim

Divide your game tokens (stones, toothpicks, small toys, etc.) into three to five piles. Each pile should have a different number of items.

Or play on paper by drawing three or more large rectangles, called "piles." Within each pile, draw several circles, called "stones." Give each pile a different number of stones.

On your turn, remove or mark out one or more stones from a single pile. You may take any number, up to the whole pile.

Whoever gets the last stone of all loses the game.

Tsyanshidzi

As in Nim, you can remove or mark out one or more stones from one pile, up to the whole pile. OR you may choose to take the *same* number of stones from *all* the piles.

Whoever gets the last stone of all loses the game.

Notes & House Rules

Nim Variations

MATH CONCEPTS: logic and strategic thinking.
PLAYERS: only two.
EQUIPMENT: pencil and paper.

One Won

No equipment required. The first player chooses a 2-digit number. On your turn, either subtract one or divide the number in half (ignore fractions). Whoever reaches the number *one* wins.

2-D Nim

Draw a rectangle and divide it into smaller squares.

On your turn, color one, two, or three of the squares—but you can only take multiple squares if they are touching sides.

The player who colors the last square loses.

2-D Tsyanshidzi

You may color one or more squares from a single row, or the *same* number of squares in *all* the rows.

Notes & House Rules

Race to the Pharaoh's Treasure

Math Concepts: logic and strategic thinking.
Players: two players or two teams.
Equipment: gameboard, 45 dried beans or other small tokens.

The Story

Your team of archaeologists have discovered a new pyramid hiding an Egyptian Pharaoh's secret stash. Unfortunately, a band of robbers found your dig site and is trying to steal the treasure.

Now it's a race to see who can remove the last stone and uncover the Pharaoh's Treasure.

Set-Up

Place one dried bean or other small token on each stone of the gameboard.

How To Play

Beginning at the side or top of the pyramid and working their way down, players take turns removing up to five stones from the pyramid (by picking off those beans).

- You must take at least one stone.
- You can dig down into the pyramid in any direction.
- Each of the stones you take must touch at least one of the others from that turn.
- Stones may stick out in a cantilever, but they can't float in the air. Every stone remaining at the end of your turn must either be on the bottom row or must touch stones that reach down to the bottom row.

The player (or team) who takes the last stone uncovers the treasure and wins the game.

A cantilever is allowed:

Floating is *not* allowed:

Race to the Pharaoh's Treasure

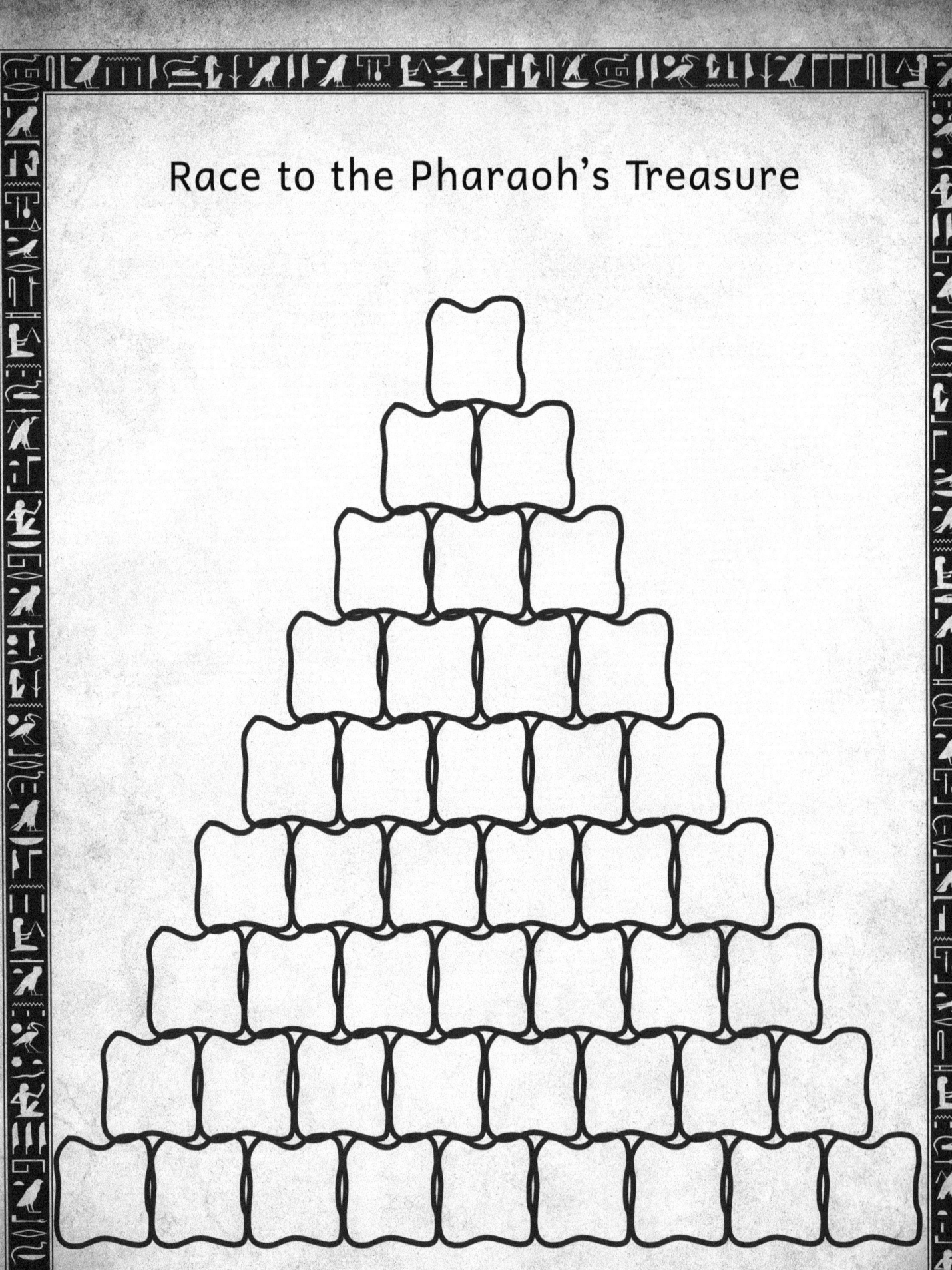

Race to the Pharaoh's Treasure

Notes & House Rules

Make Your Own Nim Game

When we reason about game strategy, that's mathematical thinking. We are dealing logically with the things we see and the facts we know.

So play one of the Nim games a few times just for fun. And then take a break and talk about the things you noticed. Wonder together about what might happen the next time you play.

- Did you figure out a winning strategy?
- How might your opponent try to block you?
- Would you rather play first or second, or does it matter?

Make Your Own Math

When you create your own games, you do mathematical reasoning at a deeper level, considering the types of choices to give your players and how limiting those options will affect the game.

First, think about a theme or story behind your game. Perhaps your players are dragons trying to hide their gold, or ninja warriors trying to sneak into an enemy camp, or birds building nests on the branches of a tree. Or would you rather make an abstract game of counting or picking up stones?

Will you stick to the tradition of a 2-player strategy game, or will you make a way for more people to join in the fun?

Decide what options players will have in your game:

- Will they be taking physical items like stones or game pieces?
- Will they be moving or claiming positions on a gameboard?
- Or will they be choosing among abstract ideas like numbers or shapes?

In a Nim game, players remove options on each turn until no choices remain. How will players remove options in your game? What choices can they make? What are the limits? How will they win (or lose) the game?

Write out the instructions for your game. Copy what you need from the games in this book, modifying the rules to fit the way you want to play. Draw your gameboard, if needed, and collect whatever game pieces you desire.

Test your new game by playing with a friend. How did it go? Do you want to tweak the rules?

Share Your Game Online

I'd love to see what you create! Submit your new game to the Student Math Makers Gallery at TabletopAcademy.net/math-makers.

Game Title: _____

Describe your game. Does it have a story?

Players' Options:

○ Giving or taking physical items

○ Moving or claiming positions on a gameboard

○ Choosing among abstract ideas like numbers or shapes

○ Something else. Explain:

Gameboard Sketch:

Write your finished game on a fresh sheet of paper.
Give it a border and other decorations to make it look cool.

Game Title: _____

Describe your game. Does it have a story?

Players' Options:

○ Giving or taking physical items

○ Moving or claiming positions on a gameboard

○ Choosing among abstract ideas like numbers or shapes

○ Something else. Explain:

Gameboard Sketch:

Write your finished game on a fresh sheet of paper.
Give it a border and other decorations to make it look cool.

TABLETOP MATH GAMES COLLECTION

EARLY MENTAL MATH

6 WAYS TO PLAY MATH WITH PRIMARY STUDENTS

Notes & House Rules

Coin Chain

MATH CONCEPTS: value of coins, logic, thinking ahead.
PLAYERS: best for two.
EQUIPMENT: ten or more assorted coins.

How To Play

Arrange the coins in a line (straight or curved) so that each coin touches two others, except the coins on the ends of the line will have only one neighbor. Or arrange them in a loop so every coin touches two others.

The youngest player takes the first turn removing a coin. If the coins form a loop, you may choose any coin. On subsequent turns, or if the coins begin in a line, you may choose the coin at either end. Keep the coins you take in a stash pile.

When all the coins are taken, players count the value of their stashes. Whoever has collected the most money wins the game.

*It's your turn. Should you take
the nickel (5 cents) or the penny (1 cent)?*

Notes & House Rules

Make and Take

Math Concepts: addition, subtraction, multistep calculation.
Players: only two.
Equipment: one deck of playing cards (face cards removed).

How To Play

Deal each player five cards. Turn the remainder of the deck face down as a draw pile. Players do not take turns, but play at the same time:

- Choose one of your cards as a challenge to your opponent. Hold it out face down. When both of you are ready, turn the challenge cards face up.
- Try to combine two or more of the cards left in your hand by adding or subtracting them to make the number on your opponent's challenge card. If you can make it, lay down those cards and say the calculation.
- If you make your opponent's card, you can take it for your score stash. If you can't make the challenge card, it goes on the discard pile.

Then pick up the cards you used to make the challenge number, and return them to your hand. Draw one more to bring your hand back to five for the next turn. If necessary, shuffle the discards to replenish the draw pile.

The first player to capture ten cards wins the game. If both players get ten on the same turn, then add up the numbers on your captured cards, and the player with the highest total wins.

Variations

For a shorter game, play to five or seven captures. For a tougher game, deal four cards to each player, so each will have only three cards left for making the challenge number.

Bonuses: If you made the challenge number using all the cards in your hand, you can capture the smallest of those cards as a bonus. You may also claim a bonus if you used cards that are all the same suit—and if your cards all matched the suit of the challenge card, you can capture the smallest two of them.

Bonuses are additive, so if you used all the cards in your hand and they all matched the challenge card's suit, you could claim three bonus cards.

Catch Up: If you can't make your opponent's challenge card on one turn, leave it on the table (don't discard). On the next turn, you may either make the new challenge card by itself or try to claim both of them by making their sum.

Notes & House Rules

For free printable gameboards, download the Number Game Printables Pack at TabletopAcademy.net/free-printables.

Tiguous

MATH CONCEPTS: addition, subtraction, multistep calculation.
PLAYERS: two or more.
EQUIPMENT: gameboard, three six-sided dice, pencil or marker(s), scratch paper for keeping score.

Set-Up

Use one of the Tiguous gameboards, or have players make their own gameboard:

- Draw a 6 × 6 grid of squares, each big enough for a two-digit number.
- The first player writes the number one in any square. The next chooses a square for the number two. Players take turns writing the numbers 1–18 anywhere they wish, one number per square.
- After eighteen, the next player goes back to one, and the turns continue until the board is full.

How To Play

On your turn, roll all three dice. Add or subtract these three numbers in a two-step equation that equals the number in any unmarked square on the gameboard.

Think of as many possible combinations as you can, in order to choose the highest-scoring square. Mark your answer on the gameboard with a large X. At the same time, say out loud how you calculated the number.

You score 1 point for the square you marked, plus 1 point for each already-marked square that is touching (*contiguous* to) any side or corner of your number's square. The maximum score for any turn is 9 points. If all the numbers you can make have already been marked, you score a zero—but if anyone else can find a valid calculation using your dice, that player may challenge you, mark the square, and steal those points.

If another player thinks you made an arithmetic mistake, that person may challenge your answer before the next player rolls the dice. If your answer was wrong, the challenger takes the points you would have won, and you score zero. If your calculation is correct, you get one bonus point for having withstood the challenge.

Play until each player has had ten turns, or five turns each for three or more players. Whoever has the highest total score wins the game.

Which square would you mark?

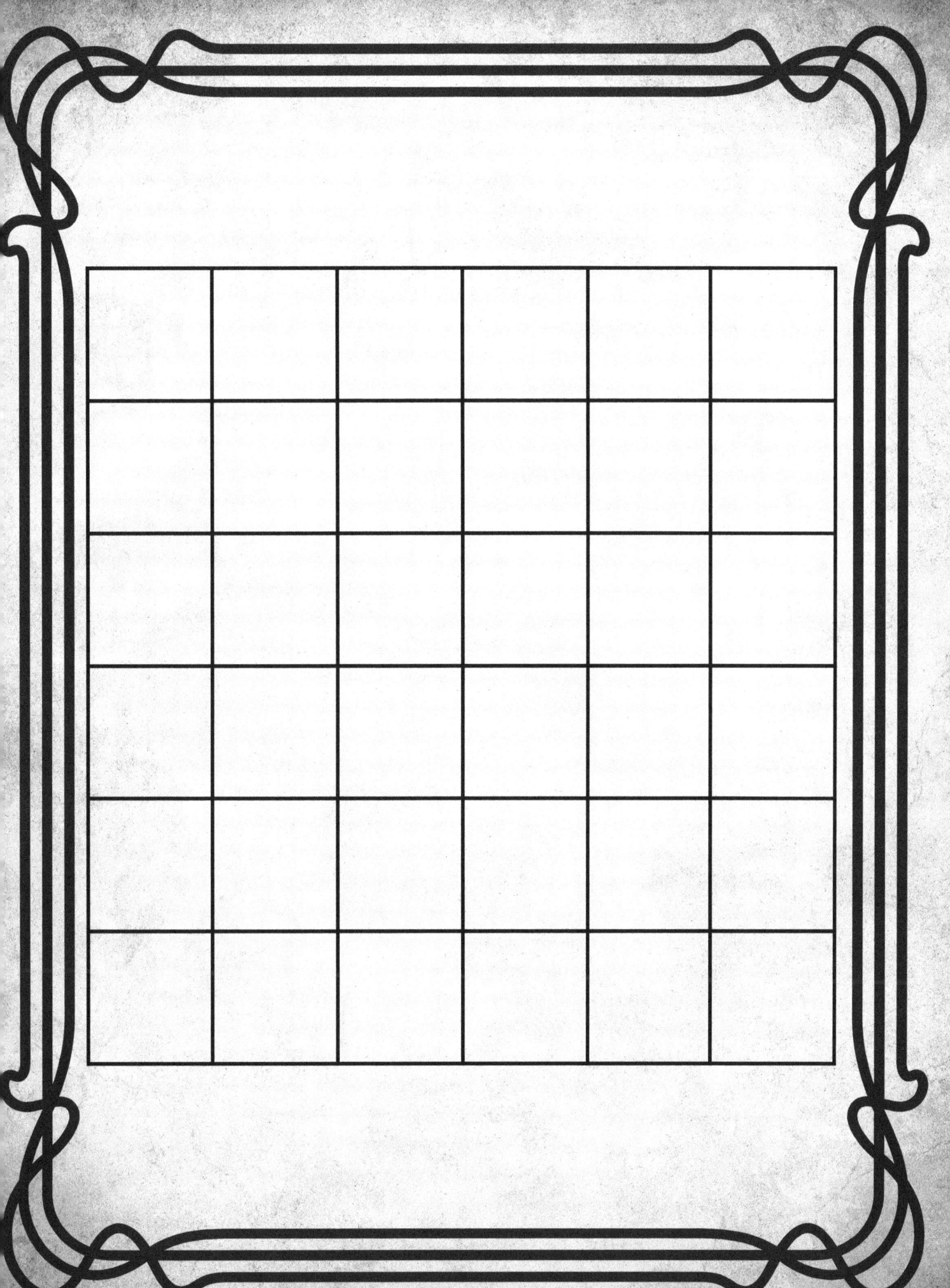

1	2	3	4	5	6
7	8	9	10	11	12
13	14	15	16	17	18
1	2	3	4	5	6
7	8	9	10	11	12
13	14	15	16	17	18

1	2	3	4	5	6
12	11	10	9	8	7
13	14	15	16	7	8
18	1	18	17	6	9
17	2	3	4	5	10
16	15	14	13	12	11

Notes & House Rules

Tiguous-Tac-Toe

MATH CONCEPTS: addition, subtraction, multistep calculation.
PLAYERS: two or more.
EQUIPMENT: gameboard, three six-sided dice, pencil or marker(s), scratch paper for keeping score.

Set-Up

Use one of the Tiguous gameboards, or have players make their own gameboard to share. Each player chooses a unique symbol to mark their squares: X, O, star, initials, or whatever you like.

How To Play

On your turn, roll all three dice. Add or subtract these three numbers in a two-step equation that equals the number in any unmarked square on the gameboard.

Think of as many possible combinations as you can, in order to choose the best-placed square. Mark your answer on the gameboard. At the same time, say out loud how you calculated the number.

The first player to get three squares in a row wins. Rows may be vertical, horizontal, or diagonal.

For a longer game, try to get four or five in a row.

Multiplayer Extended Game

With more than two players, keep going until almost all the numbers are marked. Any player who cannot mark a square three turns in a row drops out of the game.

The game ends when the last player strikes out.

Connect all your own marked squares that touch each other vertically, horizontally, or diagonally. Whichever player has the longest string of contiguous squares wins the game.

Notes & House Rules

Hit Me

MATH CONCEPTS: integer addition, absolute value.
PLAYERS: two or more.
EQUIPMENT: playing cards (two decks may be needed for a large group).

How To Play

Agree on which color represents negative numbers. Aces count as one, and face cards as ten. Choose one player as dealer.

The dealer gives each player one card face down and then turns one card face up beside each face-down card. Players do not pick up their cards. You may peek at your own face-down card as often as you like, but keep it hidden from the other players until the end of the round. The face-up card remains visible to all players.

Mentally add the numbers on your cards, taking into account both positive and negative integers. Your sum may go below zero.

When all players have had time to check their cards, the dealer asks each in turn whether they want a *hit*—an extra card dealt face up so everyone can see it. If you want the extra card, say "Hit me!" Add your new card to your running total, but don't say your sum out loud.

Last of all, the dealer may take a hit.

Then each player can ask for a second hit, and then a third, up to five hits (for a maximum of seven cards). Players may *hold*—stick with the cards they have—at any time, but they may not change their minds later and ask for a hit. The round ends when all players have either held or taken a total of seven cards.

At the end of the round, players turn their hidden cards face up and announce their scores.

The player with the lowest *absolute value*—the sum closest to zero, whether positive or negative—wins the round and becomes the new dealer. In case of a tie, the dealer hands the deck to any player who hasn't dealt recently.

Variation

Do you hate relying on luck? Add a bit of strategy to the game by allowing the ace to count as one or eleven, player's choice.

The dealer could allow each player to take his or her hits all at once, then move to the next player. But that system can feel more tedious, with too much idle time between turns.

Notes & House Rules

Bowling

MATH CONCEPTS: addition, subtraction, multistep calculation, column addition (for final score).
PLAYERS: any number.
EQUIPMENT: gameboard, one deck of playing cards (face cards removed) or three ten-sided dice, ten tokens for covering numbers, pencil or pen, and (optional) scratch paper for adding up the scores.

Set-Up

Use the Bowling gameboard, or draw your own gameboard and use the Bowling scoresheet for a larger writing space.

Place the deck face down where all players can reach.

How To Play

When it's your turn, roll your bowling ball by flipping up three cards (or rolling the dice). On your gameboard, cover the pins you knock down:

- The numbers on your cards.
- Any numbers you can make by adding or subtracting two cards.
- Any numbers you can make by adding and subtracting with all three of the cards. Each card may be used only once in each calculation.

For example, if you roll a two, four, and ace, you can cover those numbers and also $3 = 4 - 1$, $7 = 1 + 2 + 4$, and more.

If you cover all the numbers, you rolled a strike, and that's the end of your turn.

If you don't cover all the numbers, discard those cards and turn up three more for your next ball. Can you cover the remaining numbers now? If so, that's a spare.

Your score for each turn is how many numbers you cover, with special rules for strikes and spares as explained below.

After you roll two balls, the next player takes a turn. Keep going until each player has ten turns. If the draw pile runs out of cards, reshuffle the discards to replenish it.

How to Use the Score Sheet

The rules for keeping score in bowling seem confusing until you've played a few games. Each larger square box on the score sheet is called a frame and represents your score for one turn. The two (or three) smaller square boxes are for keeping track of each time you roll your ball.*

One complete game of bowling consists of ten turns, with a maximum score of 300 points. You might think that with ten numbers to cover in each turn, the most you could score would be 100 points, but strikes and spares get bonuses.

For each turn, write down how many numbers you covered with your first ball in the first small box, or if you got a strike, mark an X there. If you threw a second ball, mark down how many more numbers you covered in the second small box, or if you got a spare, write a slash (/).

If you did not get a strike or spare, add the numbers you wrote down and put that total score at the bottom of the frame.

A spare earns one bonus ball, and a strike earns two, which means you can't finish scoring your frame yet. On your next turn, the balls you throw will count not only for their own frame but also as bonus points on your strike or spare.

After you've thrown your next ball (or two balls, for a strike), add ten for the strike or spare plus the value of the bonus(es) and write in that total at the bottom of the unfinished frame.

Notice that the bonus balls get counted twice, once as part of the strike or spare and once for their own frame.

And if you throw three strikes in a row, the third one will be counted three times:

- ♦ As the second bonus throw in the first strike frame.
- ♦ As the first bonus throw on the second strike.
- ♦ And finally as part of its own frame, along with its own bonus balls.

When you get to the last frame, you will see three small boxes. That's because any players lucky enough to throw a strike or a spare still get their bonus balls. If you don't get a strike or spare, ignore that third little box.

Finally, add up all the points from all ten frames to find your total score for the game.

*For free printable score sheets, download the Number Game Printables Pack at TabletopAcademy.net/free-printables.

For More Information

Coin Chain

Alexander Bogomolny posted this coin game on his Cut the Knot website. In his interactive version, you can choose how many coins you want and play against the computer.

- ♦ cut-the-knot.org/Curriculum/Games/Coins.shtml

Make and Take

This game was created by Grand Valley State University mathematics associate professor John Golden and math/physics teacher Nicholas Smith and published on Golden's fantastic Math Hombre blog.

- ♦ mathhombre.blogspot.com/2011/11/make-and-take.html

And don't miss Golden's amazingly extensive math games linkfest:

- ♦ mathhombre.blogspot.com/p/games.html

Tiguous

Tiguous is a simplified version of F. W. Broadbent's game Contig, which is played on a larger board and allows the use of multiplication and division. I took the name from an even simpler version by Constance Kamii.

Broadbent, F. W. "Contig: A Game to Practice and Sharpen Skills and Facts in the Four Fundamental Operations," *The Arithmetic Teacher*, May 1972.

Kamii, Constance, with Sally Jones Livingston. *Young Children Continue to Reinvent Arithmetic, 3rd Grade: Implications of Piaget's Theory*, Teachers College Press, 1994.

♦ ♦ ♦

> "Be careful! There are a lot of useless games out there. Look for problem solving, the need for strategy, and math content.
>
> "The best games offer equal opportunity (or nearly so) to all your students. Games that require computational speed to be successful will disenfranchise instead of engage your students who need the game the most."
>
> —John Golden, "Math Games for Skills and Concepts"

TABLETOP MATH GAMES COLLECTION

INTERMEDIATE MENTAL MATH

6 WAYS TO PLAY MATH IN THE MIDDLE GRADES

Notes & House Rules

Distributive Dice

MATH CONCEPTS: addition, subtraction, multiplication, distributive property, rectangular area, multistep mental math.
PLAYERS: two to four.
EQUIPMENT: graph paper (1 cm squares for two players, or ¼-inch squares for more), three six-sided dice, colored markers.

How To Play

Each player will need a colored marker to shade in the gameboard squares, and the colors must be different enough to be easily distinguished because the players share a single sheet of graph paper. Each player colors a large starting dot on one corner of the grid, as far apart from the others as possible.

On your turn, roll all three dice. Choose two of the numbers to add or subtract, and that answer will form one side of your rectangle. The third die gives the other side. Draw the rectangle on the graph paper so that it shares *at least* one corner with your current territory and does not overlap squares already claimed by any player. Inside the rectangle, write the area of your newly conquered space.

The game ends when a player cannot draw any rectangle to match the dice. Players add up the areas of all their rectangles, and whoever has conquered the most territory wins.

Who do you think has the best position so far?

If you were Red and rolled these dice, what rectangle would you make?

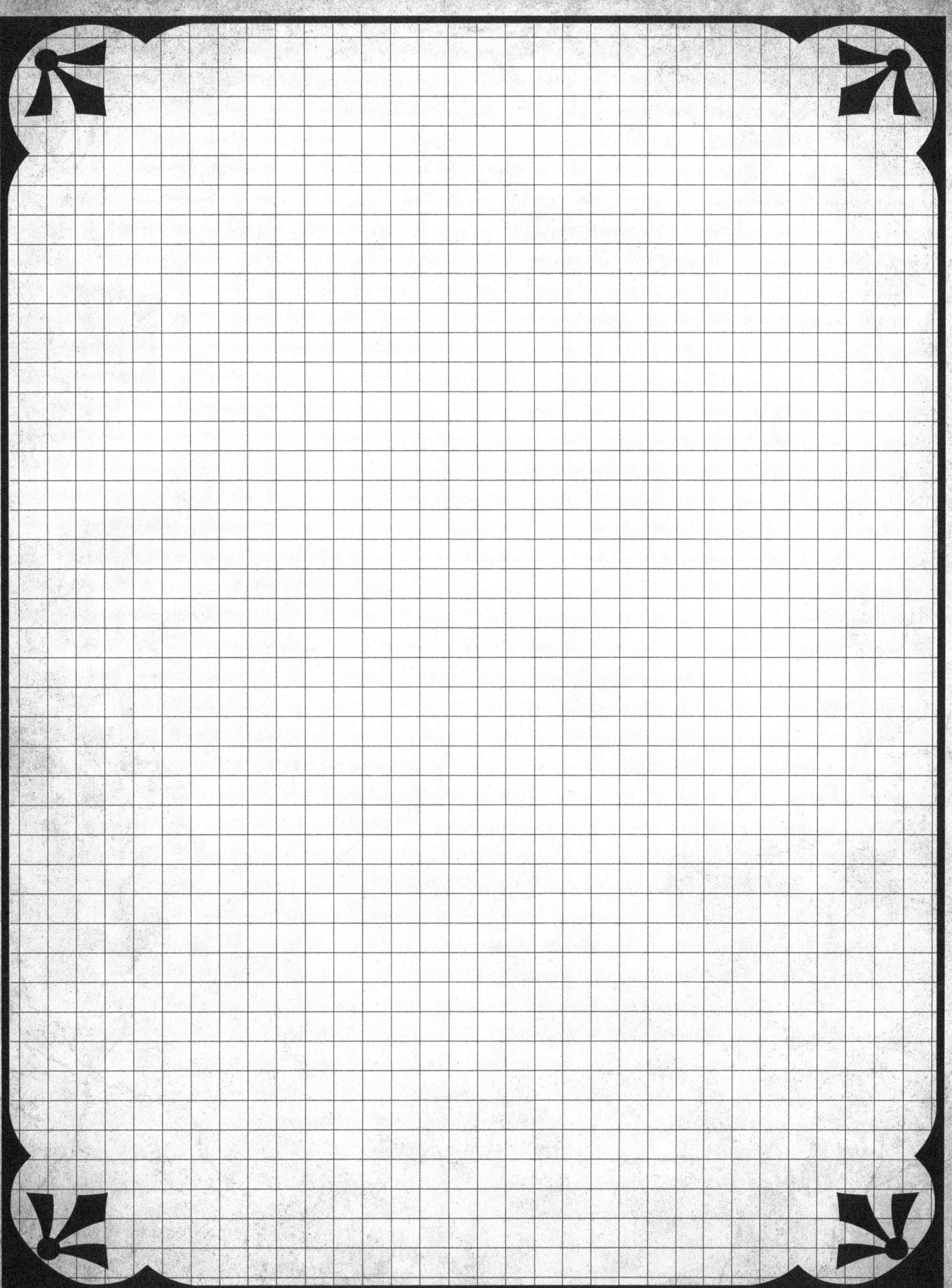

Notes & House Rules

Leapfrog

Math Concepts: addition, subtraction, multiplication, division, multistep math.
Players: any number.
Equipment: your choice of hundred chart, one regular deck of playing cards, one small toy or token per player.

How To Play

Deal four cards to each player, and place all the tokens near the first (single-digits) row of the hundred chart. Turn the rest of the deck face down as a draw pile.

On your turn, draw one card. Then choose your move:

- Use any three of the number cards in your hand to form a two-step calculation that equals any number in the next-higher row of the chart. (Face cards have no number value in this game.) Show your cards and say your equation, then jump your token to that square. Discard one of the cards you didn't use.

- Play a face card by laying it on the table in front of another player, who must move his or her token down to any space in the next-lower row. The face card counts as your discard, ending your turn.

- If you can't make a jump or play a face card, choose any card from your hand to discard.

Two or more players' tokens may share a square on the gameboard. But if your token gets moved down by a face card, you may not use the same calculation twice in succession. You may jump back up to the same square you used before only if you find a different way to calculate that number.

If the deck runs low, shuffle the discards back into the draw pile.

The first player to leap from row to row all the way to the top and then make a number greater than one hundred jumps off the chart and wins the game.

Variations

House Rule: Are the higher numbers too hard to reach? Allow players to use four cards in calculating jumps to the last four rows (numbers past 60).

Bonus Jumps: After you move, if you can use the same three cards to calculate a number in the next row, you can jump again.

Leapfrog Solitaire: Turn up six number cards. Choose any three cards to make a jump. Keep re-using the same six cards. Can you leap all the way to the end of the chart?

1	2	3	4	5	6	7	8	9	10
11	12	13	14	15	16	17	18	19	20
21	22	23	24	25	26	27	28	29	30
31	32	33	34	35	36	37	38	39	40
41	42	43	44	45	46	47	48	49	50
51	52	53	54	55	56	57	58	59	60
61	62	63	64	65	66	67	68	69	70
71	72	73	74	75	76	77	78	79	80
81	82	83	84	85	86	87	88	89	90
91	92	93	94	95	96	97	98	99	100

91	92	93	94	95	96	97	98	99	100
81	82	83	84	85	86	87	88	89	90
71	72	73	74	75	76	77	78	79	80
61	62	63	64	65	66	67	68	69	70
51	52	53	54	55	56	57	58	59	60
41	42	43	44	45	46	47	48	49	50
31	32	33	34	35	36	37	38	39	40
21	22	23	24	25	26	27	28	29	30
11	12	13	14	15	16	17	18	19	20
1	2	3	4	5	6	7	8	9	10

0	1	2	3	4	5	6	7	8	9
10	11	12	13	14	15	16	17	18	19
20	21	22	23	24	25	26	27	28	29
30	31	32	33	34	35	36	37	38	39
40	41	42	43	44	45	46	47	48	49
50	51	52	53	54	55	56	57	58	59
60	61	62	63	64	65	66	67	68	69
70	71	72	73	74	75	76	77	78	79
80	81	82	83	84	85	86	87	88	89
90	91	92	93	94	95	96	97	98	99

90	91	92	93	94	95	96	97	98	99
80	81	82	83	84	85	86	87	88	89
70	71	72	73	74	75	76	77	78	79
60	61	62	63	64	65	66	67	68	69
50	51	52	53	54	55	56	57	58	59
40	41	42	43	44	45	46	47	48	49
30	31	32	33	34	35	36	37	38	39
20	21	22	23	24	25	26	27	28	29
10	11	12	13	14	15	16	17	18	19
0	1	2	3	4	5	6	7	8	9

Notes & House Rules

Averages

Math Concepts: addition, division, elementary statistics.
Players: two or more.
Equipment: one deck of playing cards (face cards removed), pencil and paper for keeping score.

How To Play

Deal seven cards to each player. Consider these cards as your data set, and choose the statistic that will give you the highest score:

Mean = the number each data point would be, if the total were shared out equally. Add all the numbers in your set, divide that sum by seven, and then round to the nearest whole number.

Median = the middle number when the cards are arranged from least to greatest.

Mode = the number that appears most often. Not all sets have a mode.

Range = the difference between the least and greatest value. Subtract the smallest number in your hand from the largest.

Tell which statistic you are using, and add that number of points to your score. After each player has scored, pass the deck to the next dealer.

When every player has had at least one chance to deal, whoever has the highest score wins the game. Or play until someone reaches 50 points (or some other agreed-upon target).

Variation

Alternate Scoring: Award 1 point to the player with the largest of each statistic for that hand. The first player to reach 15 points wins the game.

Average Trumps: After looking at his or her cards, the player to the dealer's left chooses which statistic everyone must score for that hand. If the trump is Modes and your data set doesn't have a mode, you score zero.

Which average would you choose for this set of data?

Notes & House Rules

Twenty-Four

Math Concepts: addition, subtraction, multiplication, division, order of operations, multistep mental math.
Players: any number.
Equipment: one deck of playing cards (face cards removed).

How To Play

Deal four cards to each player, face down. The players must leave the cards face down until everyone is ready. Set the rest of the deck to one side.

At the dealer's signal, all players pick up their hands and look at the cards. Each player tries to combine all four numbers on the cards to make twenty-four (or another designated target number).

Players may add, subtract, multiply, or divide the numbers in any order, but they may not put two cards together to make a two-digit number. Each card may be used only once in the calculation.

For example, a hand of 4, 3, 7, and 9 could make:

$$(9 \times 3) - 7 + 4 = 24$$
or
$$(9 - 7) \times 3 \times 4 = 24$$
but not
$$(9 - 3) \times 4 = 24,$$
which ignores the 7 card.

This game has an element of luck. Some hands will not make twenty-four no matter how you combine the numbers. If all players seem stumped, the dealer should give each player one more card. The players may use all five cards in their hands or choose any combination of four.

When you figure out a way to make twenty-four, lay your cards face up on the table. Explain your calculation so the other players can check it.

The first player to make twenty-four using at least four cards in legal arithmetic calculations is the winner of that hand and gets to deal the next round. Or play several hands, scoring 1 point per hand, and the first player to score 6 points wins.

House Rule: Do players get frustrated by too many hands that can't make the target? Allow the ace to stand for either one or eleven, at the player's discretion. Or use the face cards as additional numbers: jack = 11, queen = 12, and king = 13.

Notes & House Rules

Twenty-Four Variations

Slap Twenty-Four

To eliminate the element of chance, deal four cards face up in the middle of the table. All players use these cards, and whoever is the first to calculate twenty-four slaps the table. The player then explains the calculation and, if it is correct, wins the hand and scores a point.

If there is no solution possible, then the first player to say "No solution" wins the point—but if another player then gives a solution, the first player gets nothing, and the second one gains 2 points.

Twenty-Four with Variables

For a faster-paced game, include the face cards as variables (wild cards). A face card in your hand may take any value that a number card might have.

Score Twenty-Four

Give each player a piece of paper and a pencil or pen, and deal eight cards face up on the table. Set a timer for 10 or 15 minutes.

Each player writes on a piece of paper as many ways as possible to combine any two or more of these numbers to make twenty-four. Each card may be used only once in each calculation.

Every valid expression scores 1 point per card used plus a 2-point uniqueness bonus if no other player wrote the same expression.

Target Number in the Car

Each player needs a clipboard (or other hard surface to write on) with paper and pencil. Players take turns naming a number between one and twenty, until there are eight numbers named. Numbers may be repeated. All players write these game numbers on their paper.

Then the driver names a target number between one and one hundred. The driver also sets a time limit for the game—perhaps until the next gas station or rest area. Players try to make the target number as many different ways as they can. Each game number may be used only once in each calculation.

Every valid expression scores 1 point per card used plus a 2-point uniqueness bonus if no other player wrote the same expression.

Notes & House Rules

*For free printable gameboards, download the Multiplication & Fraction Printables at TabletopAcademy.net/free-printables.

Contig

Math Concepts: addition, subtraction, multiplication, division, order of operations, multistep mental math.
Players: two or more.
Equipment: your choice of gameboard, three six-sided dice, pencil or marker(s), paper for keeping score.

How To Play

On your turn, roll all three dice. Use the three numbers and the basic arithmetic operations (+, −, ×, ÷) to form a two-step calculation that equals the number in any unmarked square on the gameboard. Mark your answer on the gameboard with a large X. Say out loud how you calculated the number.

You score 1 point for the square you marked, plus 1 point for each already-marked square *contiguous* to your number's square—that is, touching any side or corner. The maximum score for any turn is 9 points. If all the numbers you can make have already been marked, you score a zero—but if anyone else can find a valid calculation using your dice, that player may challenge you, mark the square, and steal those points.

When another player thinks you made an arithmetic mistake, that person may challenge your answer before the next player rolls the dice. If your answer was wrong, the challenger takes the points you would have won, and you score zero. But if your calculation is correct, you get one bonus point for having withstood the challenge.

Play until each player has had ten turns, or five turns each for more than three players. Whoever has the highest total score wins the game.

Variations

Multiplayer Extended Game: Keep playing until almost all the numbers are marked. Any player who gets a zero three turns in a row drops out of the game. When the last player gets a third strike, the game is over. There is no bonus for the last player, other than the extra turn(s).

Contig-Tac-Toe: Two players mark numbers with X and O (or other symbols of their choice), and the first player to get four squares in a row wins. Rows may be vertical, horizontal, or diagonal. For a longer game, try five in a row.

Multiplayer Contig-Tac-Toe: Play until all squares are marked or the last player strikes out. Connect all your own squares that touch each other vertically, horizontally, or diagonally. Whichever player has the longest string of contiguous squares wins the game.

1	2	3	4	5	6	7	8
9	10	11	12	13	14	15	16
17	18	19	20	21	22	23	24
25	26	27	28	29	30	31	32
33	34	35	36	37	38	39	40
41	42	44	45	48	50	54	55
60	64	66	72	75	80	90	96
100	108	120	125	144	150	180	216

1	2	3	4	5	6	7	8
28	29	30	31	32	33	34	9
27	55	60	64	66	72	35	10
26	54	125	144	150	75	36	11
25	50	120	216	180	80	37	12
24	48	108	100	96	90	38	13
23	45	44	42	41	40	39	14
22	21	20	19	18	17	16	15

For More Information

Distributive Dice

Dan Finkel invented the Damult Dice games and shared them on his Math for Love blog. John Golden modified the game, and I tweaked that to make my Distributive Dice version.

- mathforlove.com/2010/10/a-game-to-end-all-times-tables-drills-damult-dice
- mathhombre.blogspot.com/2012/04/multiplying-game-possbilities.html

Leapfrog

My version of Leapfrog is adapted from fourth-grade teacher Chris Brewer's Leap Frog Cover-up board game, which Alice Wakefield shared in *Early Childhood Number Games: Teachers Reinvent Math Instruction,* Allyn & Bacon, 1998.

Averages

This game is adapted from the M&M&M's activity in *Acing Math (One Deck at a Time)* by The Positive Engagement Project.

- teacherspayteachers.com/Product/Acing-Math-One-Deck-At-A-Time-A-Collection-of-Math-Games-3938964

Twenty-Four

The Pagat website traces the card game Twenty-Four to Shanghai in the 1960s, so the game may have originated in China. In 1988, Robert Sun created the commercial 24 Game, which uses special playing cards. The game comes in a wide range of levels, allowing students to practice topics from simple addition to fractions, decimals, and algebra.

Contig

Broadbent, F. W. "Contig: A Game to Practice and Sharpen Skills and Facts in the Four Fundamental Operations," *The Arithmetic Teacher,* May 1972.

THE FUNCTION MACHINE

A Card Game
of Algebraic Thinking

Notes & House Rules

What Is a Function?

In daily life, to *function* is to do something, and the function of an object or machine is the task it carries out. For example, a waffle maker's function is to transform batter into a deliciously toasted treat. But a *mathematical function* does not transform the input number—it simply matches each input with the defined partner output number.

Here is an example of a function rule:

Double the number and then add seven.

$$f(x) = 2x + 7$$

[Read that as "F of X equals…"]

If we give that function the input number $x = 5$, it finds the output partner $f(5) = 17$. Read this as "F of 5 equals 17."

We can write the input and output numbers as the ordered pair (5,17), joined inside a single set of parentheses. An *ordered pair* means two numbers connected in a specified order, usually (input, output) or for coordinate graphing (x, y).

The *domain* of a function is the set of allowable input numbers, and the *range* is the set of potential outputs.

When we write a function like $f(x)$, we use *variables*—letters or symbols that stand in for some other value. The variable f names the function itself. The variable inside the parentheses (x) represents the input number. Finally, the algebraic expression after the equal sign tells the rule we will use to find the matching output number.

Functions do not have to be named f, and inputs do not have to be called x. Those are the most commonly used generic variables, but you may choose any letters or symbols you like.

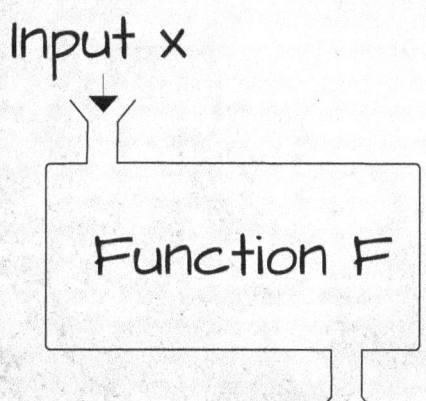

In math, a function is any rule that takes an input number and finds an output number to match it.

The Function Machine Game

MATH CONCEPTS: addition, subtraction, multiplication, division, integers, fractions, absolute value, rounding, number properties, problem solving.
PLAYERS: two or more.
EQUIPMENT: pencils and paper for keeping track of input/output numbers. Calculator optional.

Set-Up

Every player needs pencil and paper. One person takes the role of the Function Machine, making up a rule for calculating with whatever numbers the other players provide. That person should write the function rule on a hidden slip of paper for reference as needed.

All the other players make themselves an input/output chart to keep track of the game. Charts may be vertical (as shown) or horizontal.

Input	Output

How To Play

Each player in turn says a number that hasn't yet been chosen, and all players write that number in their input columns. The Function Machine calculates with that number—mentally or on scratch paper *but not aloud*—and says the matching answer. All players write that number in their output columns.

If the input number is too difficult, the Function Machine may say, "That's too hard," and ask for a different number.

Occasionally, players may give an input number that makes no sense with the chosen rule. The Function Machine can say "That number is not in my function's domain."

If you think you know the function rule, you must wait for your turn. When you give your input number, say, "I predict the output will be ___." If the Function Machine confirms your prediction, then you're allowed to guess the function rule.

The first player to identify correctly the function rule wins that round and becomes the new Function Machine.

Teaching Tips

My co-op math students have always enjoyed this game, especially when it is their turn to know the secret rule. But some kids freeze mentally at the task of creating a function rule, while others make up rules that are so complex they are nearly impossible to guess.

So I make a list of suggested function rules in advance, writing them on index cards or card-sized pieces of paper. Then I let the Function Machine players draw three cards and choose their favorite.

Here are some rules we have used:

- Add fifteen to the number:
$$f(x) = x + 15.$$

- Multiply by the next larger number:
$$f(x) = x \times (x + 1).$$

- Subtract the number from twenty-five:
$$f(x) = 25 - x.$$

- Square the number and then add one:
$$f(x) = x^2 + 1.$$

- Say the ones digit of the number.
- Add the digits in the number together.
- Say the tenths digit of the number (the first digit after the decimal place). For whole numbers, the tenths digit is zero. The Function Machine player may need a calculator to convert fractional inputs to decimal numbers.

Be warned that some function rules can be described in more than one way. For example, the rule:

Double the next higher number.

… could also be written as:

Double the number and then add two.

The player who tries to guess the rule does not have to put it in the exact words the Function Machine used, as long as the statements are equivalent.

If you are playing with younger students, it helps to have a referee who knows algebra to judge the guesses. "Double the number, and then add two" is written as $2x + 2$, while "double the next higher number" is $2(x + 1)$. Anyone who knows algebra can see these are the same.

Add fifteen to the number:

f(x) = x + 15

Subtract four from the number:

f(x) = x − 4

Multiply the number by eight:

f(x) = 8x

Cut the number in half:

f(x) = x ÷ 2

Cut the number in half and then add one:

f(x) = (x ÷ 2) + 1

Double the number and then subtract three:

f(x) = 2x − 3

Add the next larger number:

f(x) = x + (x + 1)

Multiply by the next larger number:

f(x) = x × (x + 1)

Triple the next larger number:

f(x) = 3(x + 1)

Subtract the number from twenty-five:

f(x) = 25 − x

Square the number and then add one:

f(x) = x² + 1

Divide the number by three and say the remainder.

Say the largest multiple of seven that is less than the number.

Double the odd numbers, but cut the even numbers in half.

Round off to the nearest hundred.

All odd numbers match an output of seventeen, but all even numbers match twelve.

Say the ones digit of the number.

Add the digits in the number together.

If the number is prime, say the number. If the number is not prime, say "one." (Remember that one, zero, fractions, and negative numbers are not prime.)

Say the tenths digit of the number (the first digit after the decimal place). For whole numbers, the tenths digit is zero. The Function Machine player may need a calculator to convert fractional inputs to decimal numbers.

Make your own function using multiplication:

Make your own function using fractions:

Make your own function using a percent:

Make your own function using subtraction:

Make your own two-step function:

Make your own function using division:

Make your own function using addition:

Make your own function using place value:

Make your own function using prime numbers:

Make your own function using odd and even:

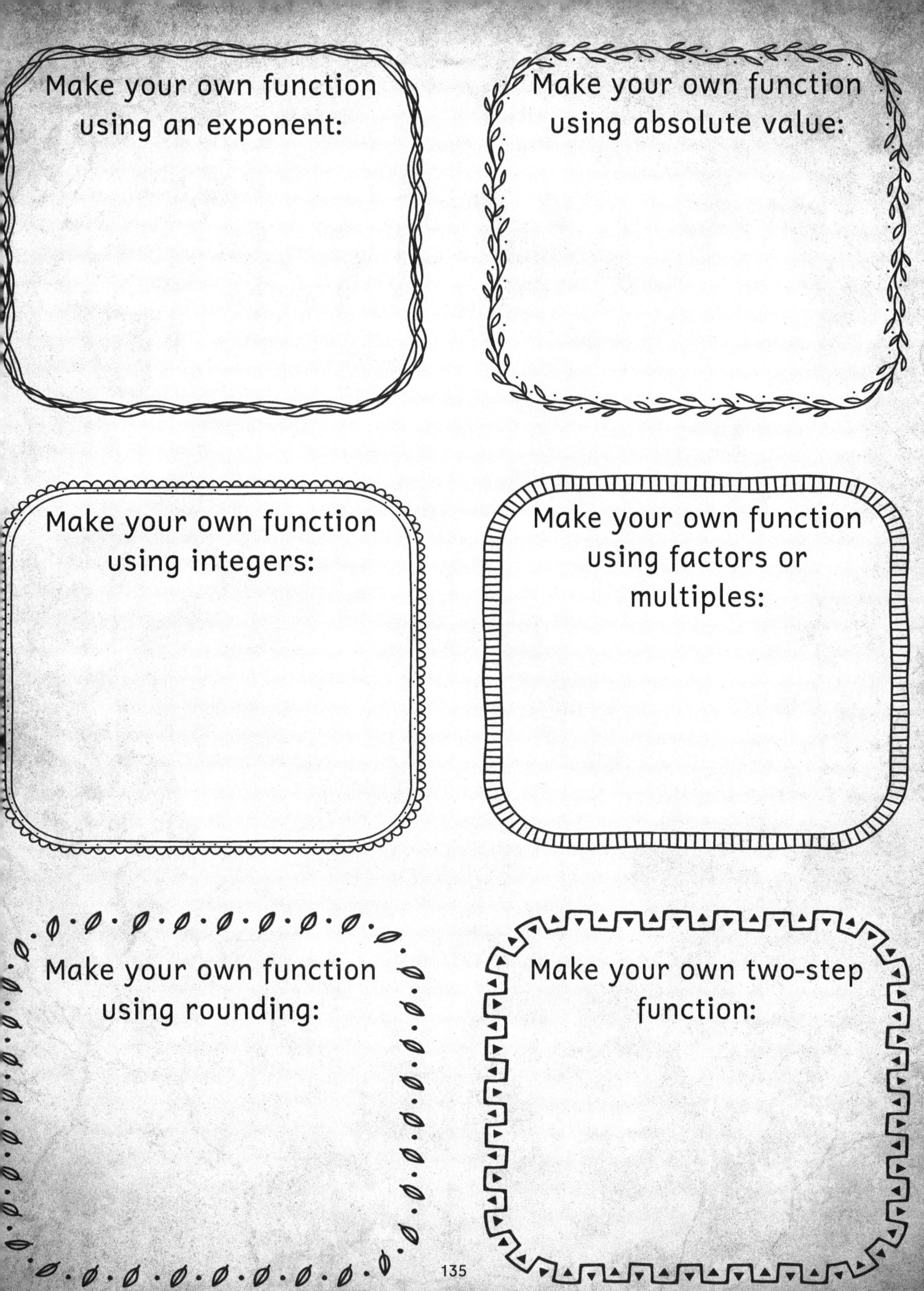

Notes & House Rules

TABLETOP MATH GAMES COLLECTION

ADVANCED MENTAL MATH

6 WAYS TO PLAY MATH WITH OLDER STUDENTS

Notes & House Rules

Masquerade

Math Concepts: addition, subtraction, multiplication, distributive property, multistep mental math.
Players: only two.
Equipment: three six-sided dice, pencils and paper for keeping score.

How To Play

On your turn, roll all three dice but keep the numbers hidden.

Add two of the numbers, and then multiply by the third. Tell your opponent the answer.

The other player says a number they think might be on one of your dice. If they are correct, reveal that die, and the guessing player scores that many points.

If two dice are the same, only show one per guess.

After three guesses, reveal any remaining unclaimed dice. Pass them to the other player, so they can roll dice for you to guess.

Play five rounds, and the highest total score wins.

Variations

Use higher-numbered dice for a greater challenge.

Damult Dice: (Any number of players.) No guessing. On your turn roll three dice. Choose two of the numbers to add together, multiply by the third, and then add that many points to your score. If you have enough dice and enough space to keep them separate, players can all roll at the same time. The first player to reach 300 points wins the game—or if two players pass 300 points in the same turn, then the highest score wins.

Double Damult: Roll six dice. Use any combination of addition, subtraction, multiplication, and division to calculate your score—but you have to use each operation at least once.

Notes & House Rules

Fight for the Center

Math Concepts: data sets, mean, median, mode, range.
Players: two or more players or teams.
Equipment: one homemade scoreboard per player/team, two decks of playing cards, pencils and scratch paper for calculations. Calculator optional.

Set-Up

Each player or team creates a scoreboard by folding a sheet of paper into quarters. Unfold the paper and write one word on each section: Mean, Median, Mode, and Range. Remove the jacks and kings from your playing card decks, but keep the queens to represent zero. Aces count as one, number cards at their face value. Agree on which color represents negative numbers.

How To Play

Shuffle the card decks together and deal four goal cards to each player or team. Players look at these cards and then place them face down on the table beside their scoreboards. You may peek at your own goal cards whenever you wish, but keep them secret from other players until you are ready to score.

Turn six cards face up on the table as the initial data set and arrange them in order from least to greatest. All players share these cards. Then deal a hand of five cards to each player or team. Place the rest of the deck face down for drawing cards. As you play, a face-up discard pile will grow next to the draw pile.

On your turn, you have two choices:

♦ Draw a card. Discard one number from the data set and replace it with a card from your hand, rearranging the set if necessary to keep the numbers in sequence. If one of your goal cards matches the new mean, mode, median, or range of the data set, turn that card face up and place it in the proper section on your scoreboard.

♦ Or switch out one of your secret goal cards. Put the goal card in your hand, and replace it with a different card from your hand. You do not draw or discard, and you may not claim a goal on this turn.

At the end of your turn, you should again have five cards in your hand.

You can only score each of the data measures once per game. If you place a card on the median, for example, your other goal cards have to fit the mode, mean, or range. The range is always a positive number, so you may take the absolute value of your goal card. For the other measures, your card must match exactly. You may only claim the mode if there are at least two copies of that number in the data set. The first player or team to score *three* of their four data measures wins the game.

Notes & House Rules

What Two Numbers?

Math Concepts: addition, multiplication, inverse operations, integers.
Players: two or more.
Equipment: no equipment needed.

How To Play

The leader chooses any two numbers and mentally figures their product and sum. Then the leader asks, "What two numbers multiply to make ___ and add up to ___?" The leader may choose any two operations to ask. For example:

- What two numbers add up to 15 and multiply to make 50? (5 and 10)
- What two numbers have a difference of 2 and a sum of zero? (−1 and 1)
- What two numbers have a product of $\frac{1}{6}$ and also have a difference of $\frac{1}{6}$? (−$\frac{1}{3}$ and −$\frac{1}{2}$) or ($\frac{1}{3}$ and $\frac{1}{2}$)

The other players race to find the numbers. The first player to name them correctly gets to lead the next round. Or with two players, just take turns trying to stump each other.

Remember to consider both positive and negative numbers when creating your puzzle. Or make it extra tricky with fractions or decimal numbers.

Notes & House Rules

*For free printable gameboards, download the Prealgebra & Geometry Printables at TabletopAcademy.net/free-printables.

Greater Than

MATH CONCEPTS: integer addition, integer multiplication, inequalities.
PLAYERS: two players or two teams.
EQUIPMENT: gameboard, one deck of playing cards, pencils or markers.

Set-Up

Players share a single gameboard. Choose a player to deal the first hand. Agree on which color represents negative numbers. Face cards count as ten, aces as eleven. The suits also represent operations:

- Spades and hearts = add the number.
- Clubs and diamonds = multiply by the number.

How To Play

Deal four cards to each player or team. Players each choose one card as their initial value, holding the card face down in front of them. At the dealer's signal, both players reveal their cards. If the two cards played have the same value, players return them to their hands and choose a different starter card.

Write the players' initial numbers in their columns. In the middle column, draw an inequality symbol (< or >) with its open end toward the greater value.

The non-dealer plays first. Choose one card from your hand and lay it on the table. Write the operation represented by that card in the first box of the next line on the gameboard (or beside the next line of the scoresheet). Then do the indicated calculation:

- For a spade or heart, add the value to each player's current score.
- For a club or diamond, multiply each player's score by that number.

Write the new sum or product in each player's column on the scoresheet. Finally, draw the correct inequality symbol in the middle column.

A hand consists of two turns for each player, beginning with the non-dealer. So the dealer takes the next turn, playing a card, writing it at the start of the third line, and doing the calculations based on the values in the previous line. The non-dealer fills the next line, and the dealer plays the final line of that hand. Players do not draw new cards after each turn.

After the dealer's second turn, whoever has the greater value wins that hand. Circle the winning score. Mix all cards back into the deck and pass it to the other player to deal the next hand.

The first player to win three hands wins the game.

Greater Than Sample Game

Tony challenges Steve to a game of Greater Than. Tony deals the first hand, and the players reveal their initial cards. Tony plays the queen of hearts, for a value of −10. Steve has the eight of spades, for a value of +8.

As the non-dealer, Steve goes first. He plays the three of clubs and writes "×3" on the next row. Then he multiplies both scores by three and writes each product in that player's column. He writes the less-than sign "<" to show that he's in the lead.

Tony lays down the ten of spades, adding 10 to each score. Steve chooses the five of hearts, adding −5 to the scores. Steve still has the greater value, but Tony gets one more turn.

Tony plays the two of diamonds, which multiplies both scores by −2. Multiplying by a negative number changes the sign of the scores, and Tony wins the hand with 50 points.

	Tony	?	Steve
	−10	<	8
×3	−30	<	24
+10	−20	<	34
+(−5)	−25	<	29
×(−2)	(50)	>	−58

♠♥ ➜ add (+)

♣♦ ➜ multiply (×)

HAND	+/×	NAME:	>?<	NAME:
1	INITIAL VALUES			
2	INITIAL VALUES			
3	INITIAL VALUES			
4	INITIAL VALUES			
5	INITIAL VALUES			

Notes & House Rules

*For free printable gameboards, download the Prealgebra & Geometry Printables at TabletopAcademy.net/free-printables.

Power Up

Math Concepts: powers (exponents).
Players: only two.
Equipment: one character sheet for each player, two six-sided dice, pencils or markers. Calculator optional.

Set-Up

Students create their own superheroes or supervillains using character sheets or regular notebook paper. Give your character a name and three superpowers, magical abilities, or technological gadgets. Describe the character's powers:

Greater Ability = ...
Lesser Ability = ...
Random Ability = ...

Briefly explain your character's origin story and draw a picture (optional).

How To Play

Each battle consists of three rounds, one for each of the superpowers. In the first round, players may choose which ability they want to use. For the second round, choose either unused power. Finally, the players call upon their remaining abilities.

Each player rolls one or two dice, as needed for their chosen superpower:

- ♦ Greater Ability: Roll one die. Use 4 as the base, your roll as the exponent.

- ♦ Lesser Ability: Roll one die. Use 3 as the base, your roll as the exponent.

- ♦ Random Ability: Roll two dice. Choose which is the base and which is the exponent.

Calculate how much damage your attack inflicts upon your opponent. Whoever does the most damage wins that round.

The player who wins two of the three rounds wins the battle. If there is a tie, fight one more round, with both players using their random ability.

Variation

Power Team Assemble: Two teams of players battle each other. For each round, teammates multiply their individual scores to calculate the total damage they inflict.

Character Name:

Greater Ability: _____
(Base 4)

Lesser Ability: _____
(Base 3)

Random Ability: _____
(Roll 2 dice. Choose one for the base, the other as the exponent.)

Your powers might be something like super speed, super strength, telepathy, telekinesis, stretchiness, fire, ice, lightning, earthquake, creature power, magic, ring/gem of power, blaster, high tech gadgets, mutant ability, alien artifact, etc.

ORIGIN STORY: How did you get your power? Are you a hero or a villain?

Character Name:

Greater Ability: _____
(Base 4)

Lesser Ability: _____
(Base 3)

Random Ability: _____
(Roll 2 dice. Choose one for the base, the other as the exponent.)

Your powers might be something like super speed, super strength, telepathy, telekinesis, stretchiness, fire, ice, lightning, earthquake, creature power, magic, ring/gem of power, blaster, high tech gadgets, mutant ability, alien artifact, etc.

ORIGIN STORY: How did you get your power? Are you a hero or a villain?

Notes & House Rules

Exponent Number Train

MATH CONCEPTS: powers (exponents), estimation.
PLAYERS: two to four.
EQUIPMENT: one set of dominoes. Calculator optional.

How To Play

Turn all the domino tiles face down on the table and mix them around to make the woodpile. Each player draws six tiles from the woodpile but does not look at them. Players arrange their tiles in a row (train), as shown below. When all players are ready, turn the tiles in your train face up without changing which side is at the top.

Estimate the value of the exponential expression represented by each tile. The bottom number (closest to you) is the base, and the top is the exponent. Your goal is to make the values in your number train increase from left to right, but of course it will be mixed up to start with.

On your turn, draw one tile and decide which way you want to turn it to create an exponential expression with the numbers on the two halves of the domino. Then you have three choices:

- Use the new tile to replace one of the domino tiles in your train. Then discard the old one, mixing it into the woodpile.
- If you don't want to use the new tile, put it back and mix up the woodpile.
- Instead of drawing a new tile, you may use your turn to rotate one of your current tiles, switching the base and exponent.

The first player to complete a train that increases in value from left to right wins the game.

Variations

Play with a different number of tiles. Try a shorter train for a quick game, or longer for a greater challenge.

HOUSE RULE: Decide how strict you will be about the "increases from left to right" rule and repeated numbers. Do you consider equivalent values (such as 1^6 and 3^0, or 4^2 and 2^4) as part of a valid train? Or must the player keep trying for a domino to replace one of the equivalents?

A domino number train, ready to flip and play.

For More Information

Masquerade

Dan Finkel invented the Damult Dice games and shared them on his Math for Love blog. John Golden played the game with a fifth-grade class, and the students created several more variations, including Masquerade.

- mathforlove.com/2010/10/a-game-to-end-all-times-tables-drills-damult-dice
- mathhombre.blogspot.com/2012/04/multiplying-game-possbilities.html

Fight for the Center

High school math teacher James Cleveland created this game and shared it on his blog, The Roots of the Equation.

- rootsoftheequation.wordpress.com/2015/07/27/fighting-for-the-center

Greater Than, Power Up, and More

John Golden trains future math teachers as an associate professor at Grand Valley State University, and trains the rest of us through the posts on his blog. I love his Math × Art webpage!

- faculty.gvsu.edu/goldenj/GameshandoutHS.pdf
- mathhombre.blogspot.com/p/games.html
- mathhombre.blogspot.com/p/mathart.html
- mathhombre.blogspot.com/2012/09/greater-than.html
- mathhombre.blogspot.com/2011/05/power-up.html

♦ ♦ ♦

"Note that it's okay for heroes to fight heroes, as those kinds of misunderstandings happen all the time. Villains might fight another villain because they're just so darn evil. (Or misunderstood.)

"Also, it's okay to end in a tie: 'We may have fought to a standstill, but I'll get you next time, Snake Lady!'"

—John Golden, "Power Up"

TABLETOP MATH GAMES COLLECTION

GRAPHING GAMES

6 WAYS TO PLAY MATH WITH OLDER STUDENTS

Notes & House Rules

*For free printable gameboards, download the Prealgebra & Geometry Printables at TabletopAcademy.net/free-printables.

Hidden Hexagons

MATH CONCEPTS: coordinate graphing (first quadrant), simple linear equations, irregular polygons.
PLAYERS: two players or two teams.
EQUIPMENT: gameboard or square grid paper for each player, pencils, ruler or other straightedge.

Set-Up

To make your own gameboards, give each player a sheet of square grid paper. Players outline two 10 × 10 grids and label the x and y axes 0–10. Label one grid Top Secret and the other grid Clues.

On your secret grid, draw a closed shape with six straight sides, with vertices on the grid intersection points (the points with whole-number coordinates) and at least one grid point entirely inside the shape. Use a ruler or straightedge for precise lines. Don't let the sides of your shape cross each other. This is your hidden hexagon. Record the ordered pairs for each vertex beside the grid.

How To Play

Players alternate turns asking for information so they can guess their opponent's secret shape. On your turn, name an equation for a horizontal, vertical, or diagonal line. For example:

- The equation $x = 3$ represents the vertical line that includes such points as (3, 0) and (3, 5).

- The equation $y = 10$ indicates the top horizontal line on the gameboard, including points like (1, 10) and (7, 10).

- The equation $y = x$ is the diagonal line from the origin slanting up at a 45° angle. It passes through such points as (2, 2) and (9, 9).

Your opponent checks that line against their hidden hexagon and tells whether each grid intersection point is a vertex, on the perimeter, inside the shape, or outside.

After you discover all six vertices of the hidden hexagon, you may guess right away. Or you may keep playing to collect more information on the secret shape.

When you're ready, connect the vertices on your Clues grid. Make sure the shape fits properly around all the marked points. Reveal your drawing and compare it to the hexagon on your opponent's gameboard.

The first player to guess ends the game.

If your drawing matches the hidden hexagon, you win the game. But if your guess is wrong on any point, the other player wins.

Hidden Hexagons Example

Imagine you drew the hexagon shown below, and your opponent asks for the y = 3 clues.

Point (2, 3) is inside the shape, and (8, 3) is a vertex.

All the other points are outside your hexagon.

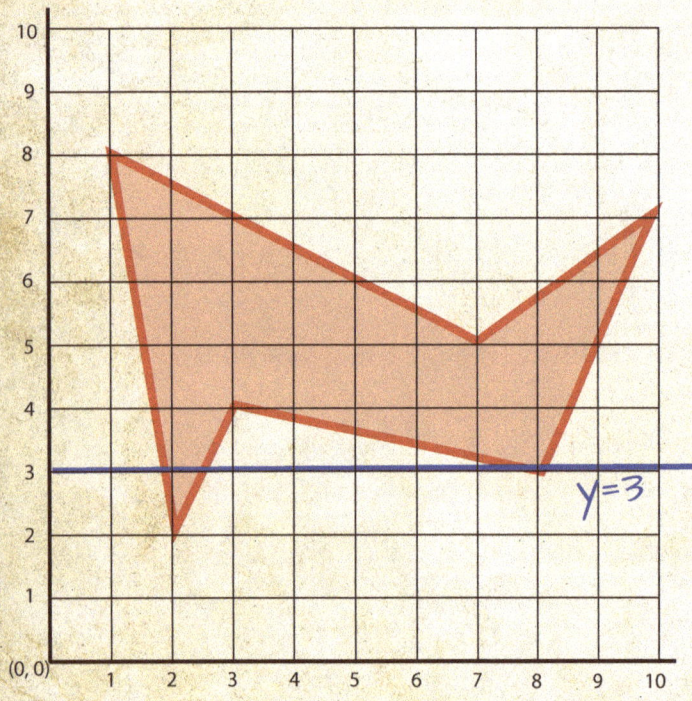

TOP SECRET:

vertex = (__1__ , __8__)

vertex = (__2__ , __2__)

vertex = (__3__ , __4__)

vertex = (__8__ , __3__)

vertex = (__10__ , __7__)

vertex = (__7__ , __5__)

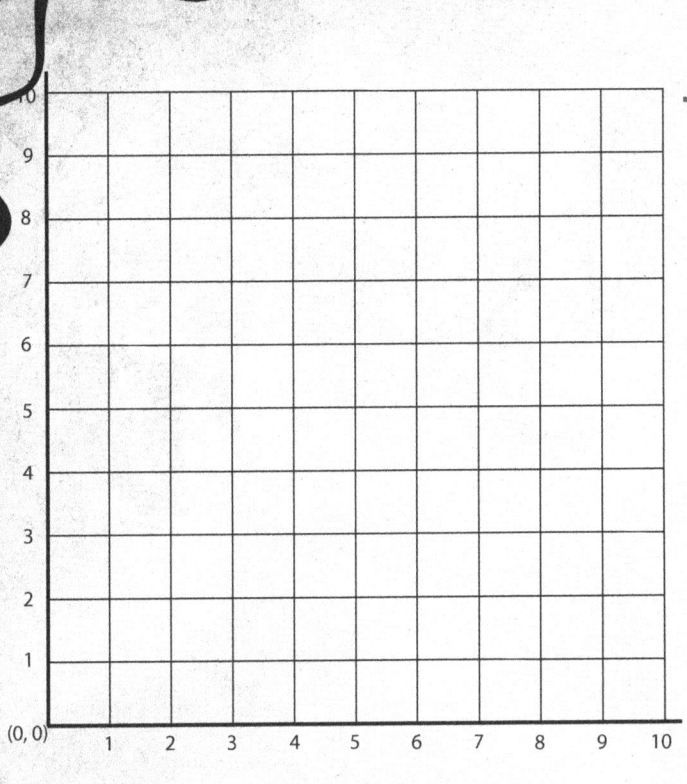

TOP SECRET!

vertex = (_____ , _____)

vertex = (_____ , _____)

vertex = (_____ , _____)

vertex = (_____ , _____)

vertex = (_____ , _____)

vertex = (_____ , _____)

(Fold the paper to hide your secret hexagon.)

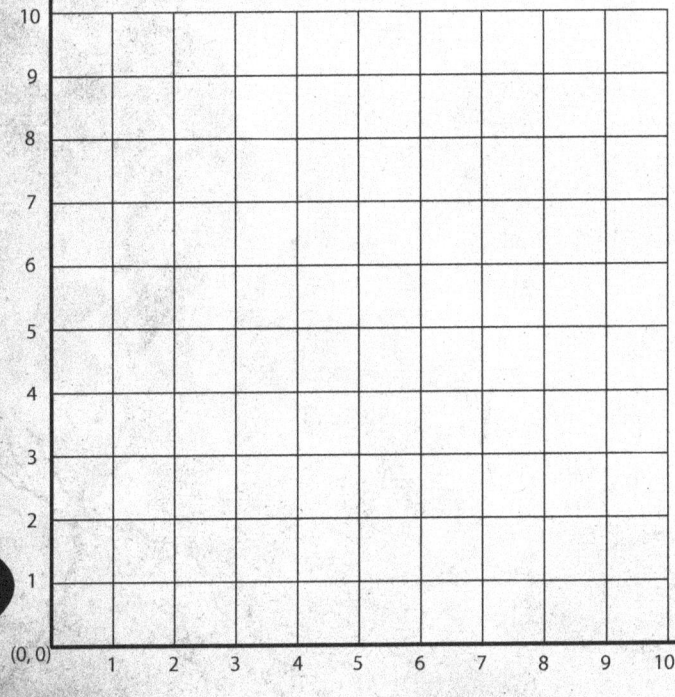

CLUES:

vertex = (_____ , _____)

vertex = (_____ , _____)

vertex = (_____ , _____)

vertex = (_____ , _____)

vertex = (_____ , _____)

vertex = (_____ , _____)

Notes & House Rules

Coordinate Gomoku

MATH CONCEPTS: ordered pairs, coordinate graphing (four quadrants).
PLAYERS: two players or two teams.
EQUIPMENT: dotty or lined square grid paper, different colored pencils or markers.

Set-Up

Players share one sheet of dotty or lined grid paper. Each player needs a different colored pencil or marker, and they may want to choose a symbol like X, O, star, or small triangle to make their marks perfectly distinct.

How To Play

The first player claims any dot on the grid by marking it with his or her symbol. If you are using lined graph paper, choose any place where two lines intersect.

This is the *origin* $(0, 0)$ for your game. Draw the horizontal and vertical lines (the x and y axes) that meet at that point.

On succeeding turns, you must name the coordinates of a point before you mark it. If you make a mistake naming the point, your opponent can give the correct (x, y) coordinates and mark it with their own symbol—and they still get to take their own turn.

The first player or team to mark five points in a horizontal, vertical, or diagonal row with no gaps wins the game.

Variations

HOUSE RULE: How do you want to handle overlines that have six or more points in a row? In traditional Gomoku, only an exact five-in-a-row line can win.

SWAP2: Does the first player win too often? What you need is a variation on the *pie rule*—one person slices the pie, and the other gets first choice of piece. Reduce the first-player advantage with these starting moves, which are often used in tournament play. The first player marks two Xs and one O. The second player selects one of three options:

- ♦ Accept these moves and mark an O. Turns continue with the first player marking X.

- ♦ Take over the Xs, so the first player goes next with O.

  - ♦ Or mark one additional X and O, then let the first player choose which letter to claim. Whoever plays O takes the next turn.

Notes & House Rules

Linear War

MATH CONCEPTS: coordinate graphing, linear equations, slope, intercepts.
PLAYERS: two to four.
EQUIPMENT: handmade game cards, pencils or markers, ruler or other straightedge.

Set-Up

Players each create a personal deck of eleven game cards. Use card stock or sturdy paper for best results.

Draw coordinate axes with the origin at the center of each card. Number the points from −5 to +5 on each axis. On each card, draw a line that passes through at least two points with integer coordinates, such as (−3, 1) and (0, 5). Mark these points with dots on the line.

CLAIM YOUR DECK: Mark the back of each card with your insignia (initials, emoticon, math symbol, etc.—your choice). Decorate as desired, so long as it doesn't bleed through the paper and obscure your lines.

How To Play

Place your game card deck face down in front of you and draw the top two cards for your hand. The player whose turn it is names the trump:

- Least slope or greatest slope
- Least or greatest x-intercept
- Least or greatest y-intercept

Each player chooses a card from their hand and holds it face down. When all players are ready, turn the cards up, and the card that best fits the trump captures them all. Each player keeps a pile of prisoner cards.

In case of a tie, leave the cards on the table for the winner of the next trick.

Players draw new cards to replenish their hands. When the decks run out, continue playing until all the cards are taken or the last round ends in a tie.

Whoever captures the most prisoners wins the game. Give the prisoner cards back to their owner, unless you are playing the Spoils of War variation.

SPOILS OF WAR: If you win the game, you may claim one card from your prisoner pile to replace a card in your deck. Erase the identification and add your own symbol on the back. Remove one card from your original deck, which the losing player may claim as a consolation prize if they wish.

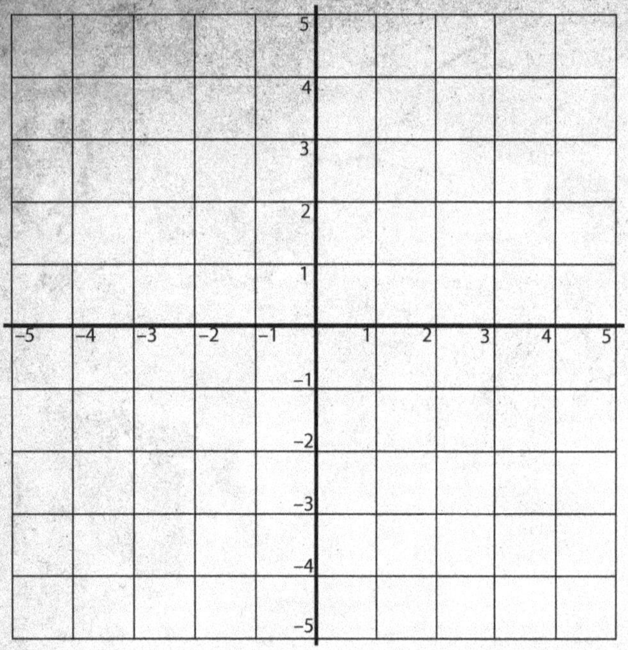
5 4 3 2 1 –1 –2 –3 –4 –5
–5 –4 –3 –2 –1 1 2 3 4 5

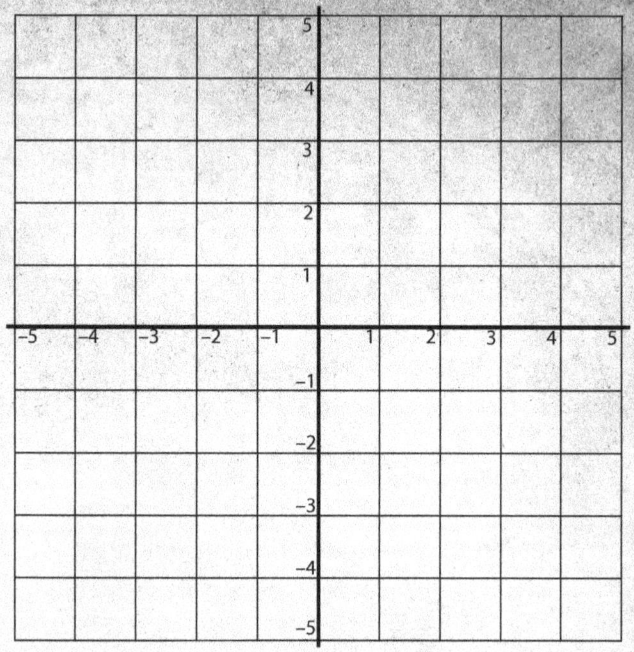
5 4 3 2 1 –1 –2 –3 –4 –5
–5 –4 –3 –2 –1 1 2 3 4 5

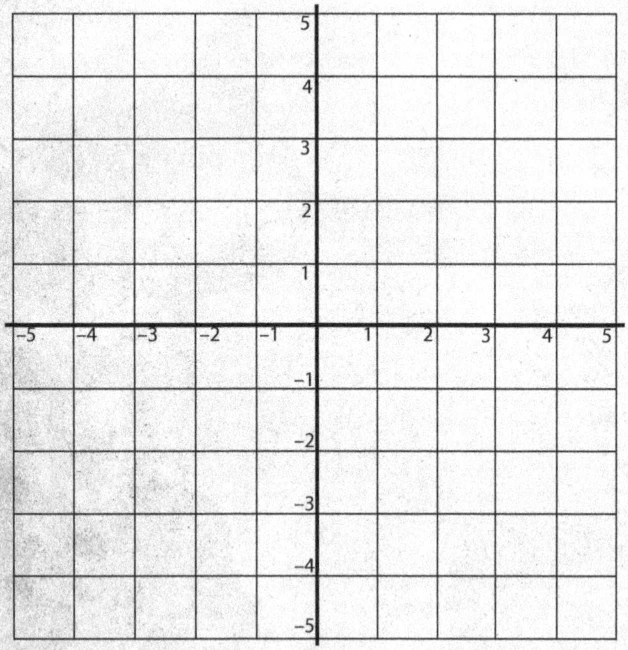
5 4 3 2 1 –1 –2 –3 –4 –5
–5 –4 –3 –2 –1 1 2 3 4 5

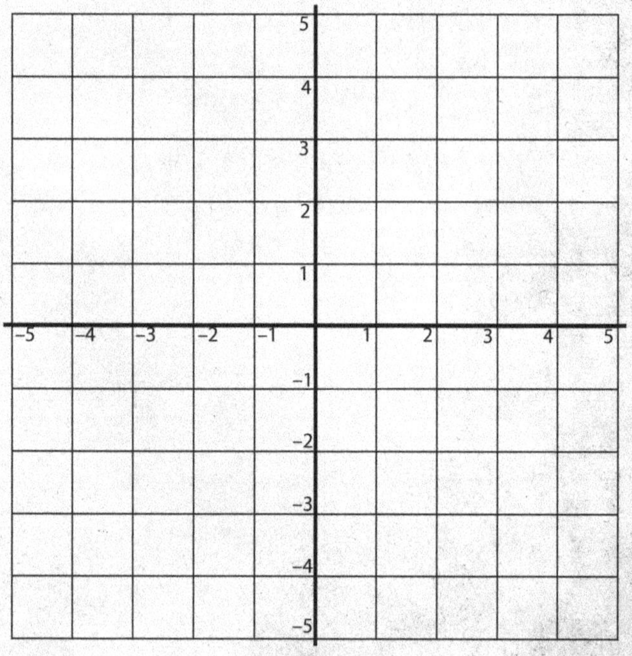
5 4 3 2 1 –1 –2 –3 –4 –5
–5 –4 –3 –2 –1 1 2 3 4 5

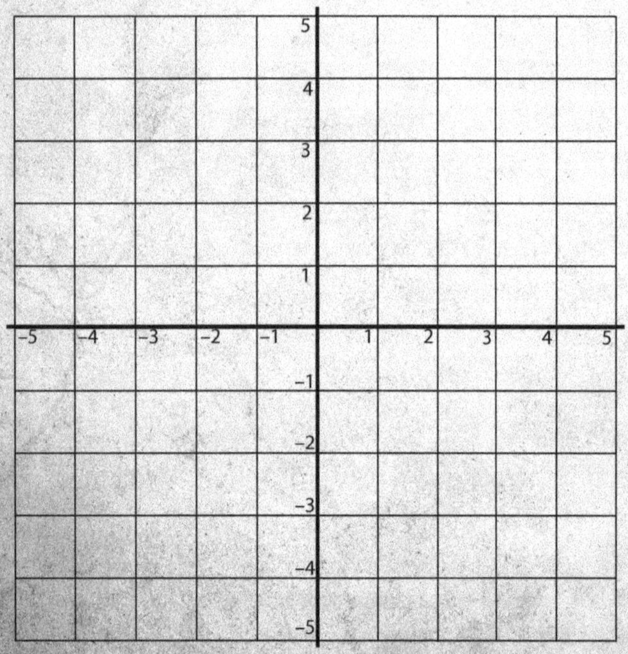
5 4 3 2 1 –1 –2 –3 –4 –5
–5 –4 –3 –2 –1 1 2 3 4 5

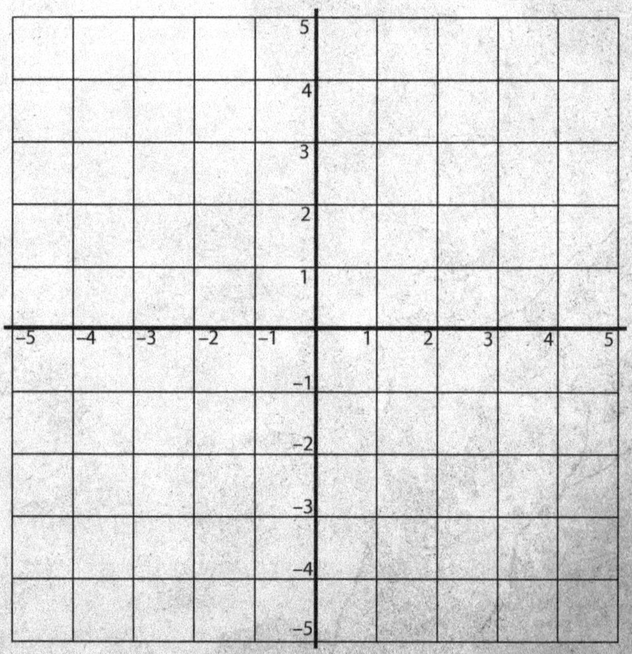
5 4 3 2 1 –1 –2 –3 –4 –5
–5 –4 –3 –2 –1 1 2 3 4 5

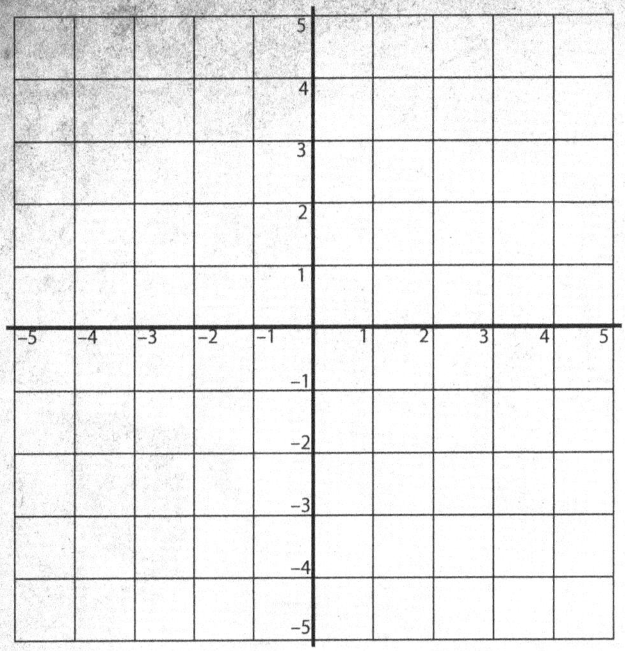
5
4
3
2
1
–5 –4 –3 –2 –1 1 2 3 4 5
–1
–2
–3
–4
–5

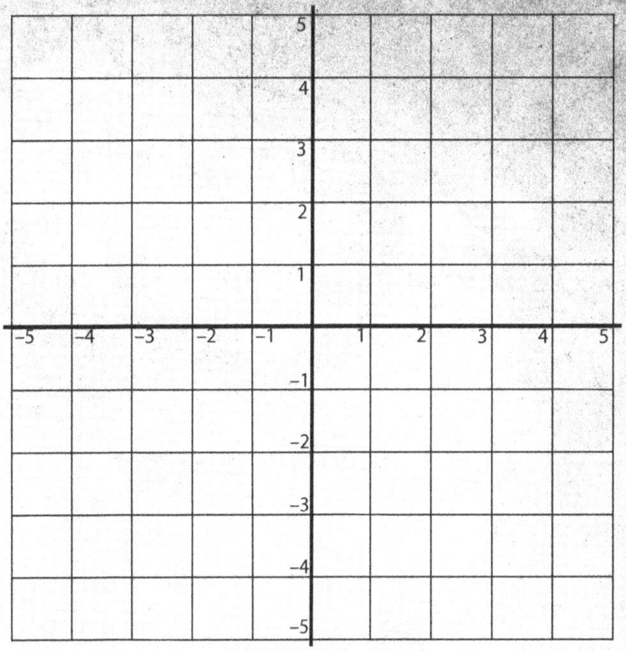
5
4
3
2
1
–5 –4 –3 –2 –1 1 2 3 4 5
–1
–2
–3
–4
–5

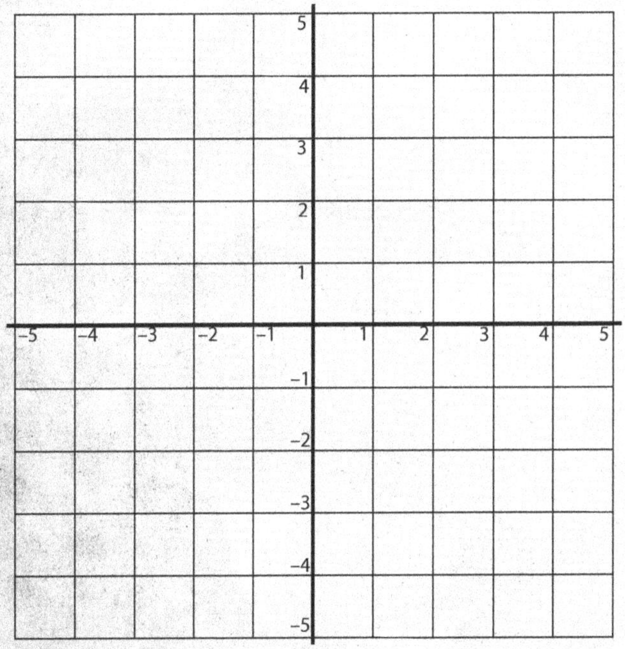
5
4
3
2
1
–5 –4 –3 –2 –1 1 2 3 4 5
–1
–2
–3
–4
–5

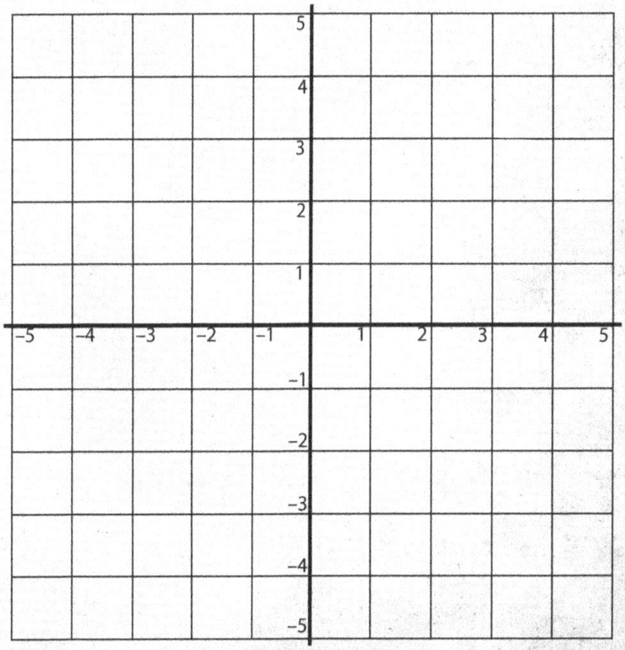
5
4
3
2
1
–5 –4 –3 –2 –1 1 2 3 4 5
–1
–2
–3
–4
–5

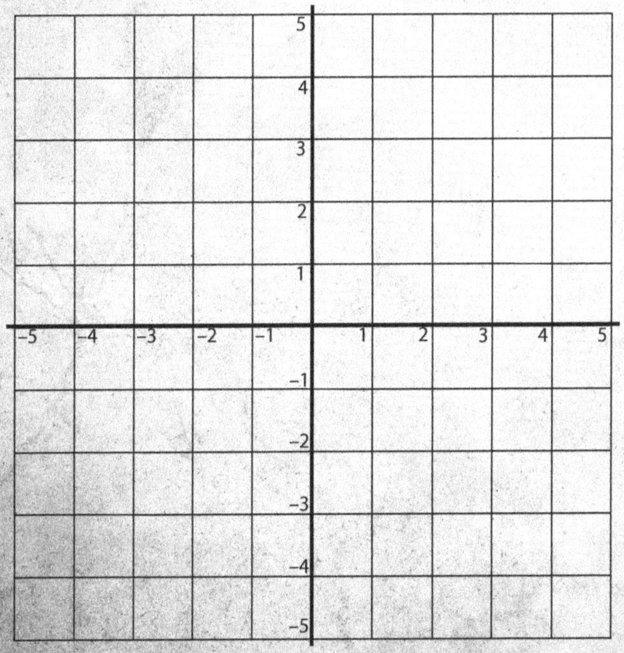
5
4
3
2
1
–5 –4 –3 –2 –1 1 2 3 4 5
–1
–2
–3
–4
–5

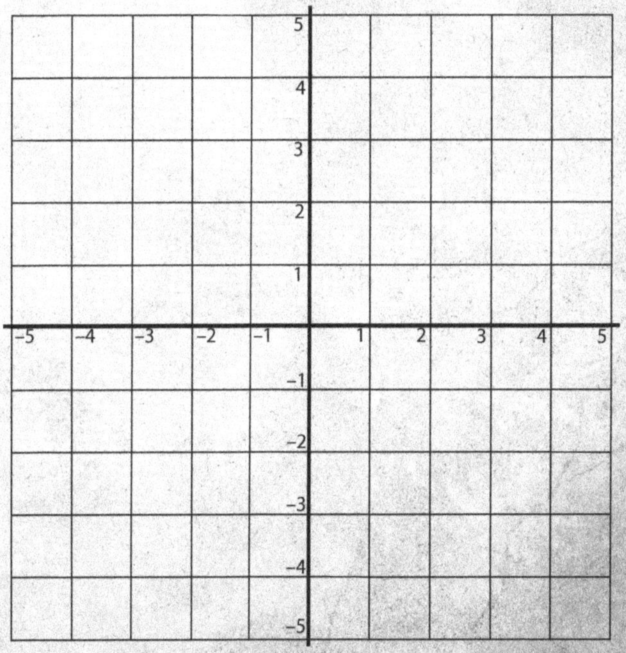
5
4
3
2
1
–5 –4 –3 –2 –1 1 2 3 4 5
–1
–2
–3
–4
–5

Notes & House Rules

Radar

MATH CONCEPTS: angle measures, polar coordinates.
PLAYERS: two to four.
EQUIPMENT: gameboard or polar (circular) graph paper, pencils or markers.

Set-Up

Choose a gameboard in degrees or radians. Each Radar gameboard has room to play four games.

Or make your own on polar graph paper. To create your own gameboard, mark circles of radius 1, 2, 3, etc. The number of circles must be equal to or greater than the number of people who want to play.

Around each circle, draw lines at convenient angles such as 30°, 60°, 90°, and so on. Or use radians: π/6, π/3, π/2, and so on.

Finally, draw an X or poison symbol in the part of the graph farthest from the origin, just below the θ = 0 axis. Whoever is forced to take that section of the graph loses the game.

How To Play

Imagine each move as a radar detector sweeping around the graph counterclockwise from the θ = 0 axis. You can move to any point where an angle line intersects a circle.

The first player chooses any intersection point on the graph, stating the move in polar coordinates (r, θ). Shade in the wedge-shaped sector swept out by that move.

On succeeding turns, players must increase either the radius or the angle, or both. Shade in the additional sections covered by each move.

A player who increases the angle may decrease the radius, if desired. Or a player who increases the radius may decrease the angle. But the radius can never go below r = 1 and the angle can never go less than θ = 30° or π/6. (Or the smallest angle marked on your circle.)

Players must shade in at least one section of the graph on every turn.

The player forced to color the last section of the graph, completing the circle at the maximum radius, loses the game.

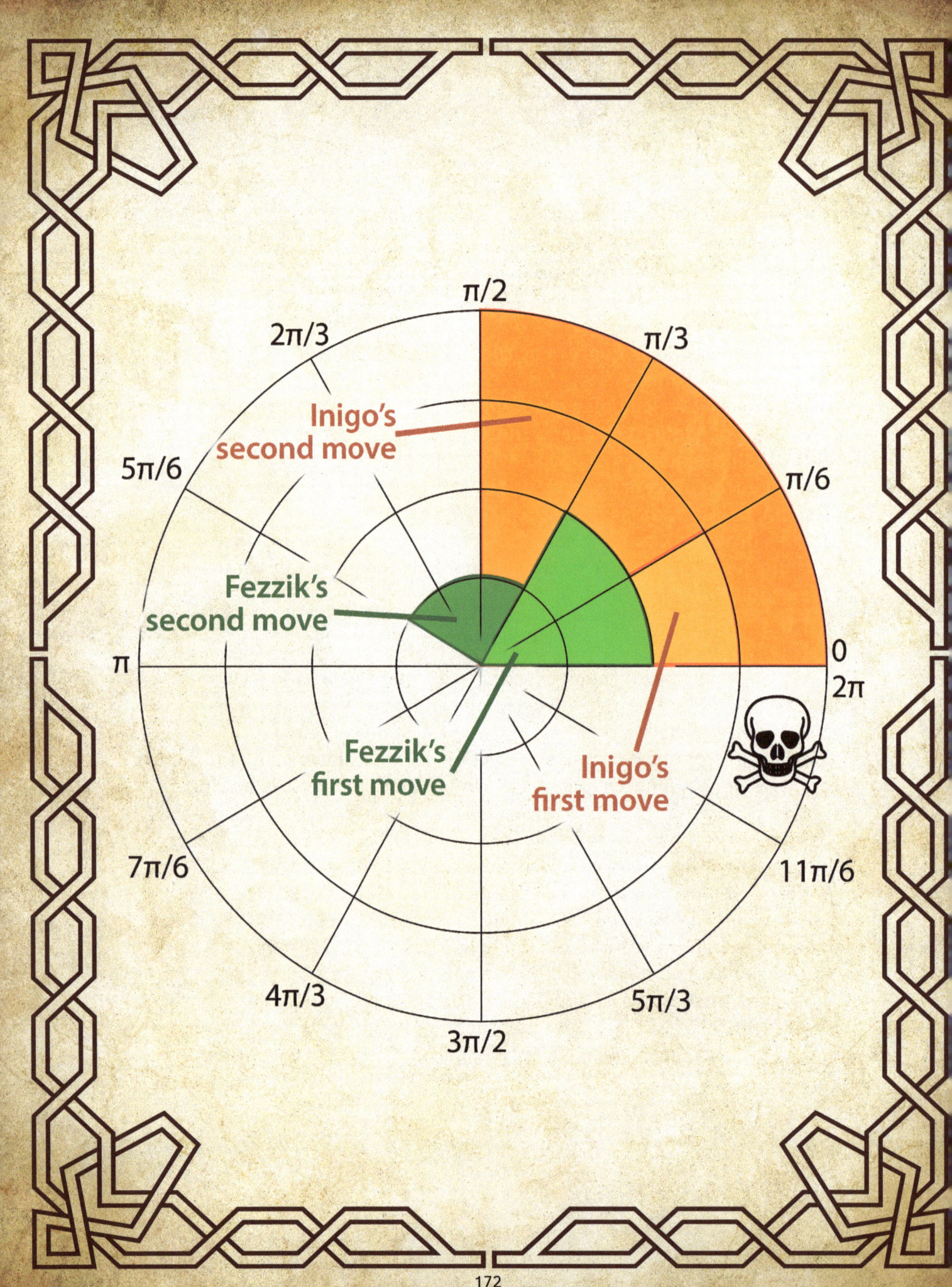

Radar Sample Game

Fezzik challenges Inigo to a game of Radar. They choose a gameboard with radius 4 and angles measured in radians.

First Round

Fezzik moves first, claiming the point $(2, \pi/3)$. He colors in the wedge from that point to the origin and down toward the $\theta = 0$ starting axis.

Inigo takes $(3, \pi/6)$, increasing the radius but going for a smaller angle. He colors his sector down to the axis and inward until it meets the already-shaded area.

Second Round

Fezzik chooses $(1, 5\pi/6)$, almost halfway around the smallest circle. He shades the wedge between that point and his previous move.

Inigo gets impatient and takes a big chunk. He marks $(4, \pi/2)$ and colors in the rest of the first quadrant.

Onward Around the Board

The shaded area will continue growing until one player has no choice but to mark the point $(4, 2\pi)$ and take the poison sector. Who do you think will get stuck with the poison?

What strategy would you use to win the game?

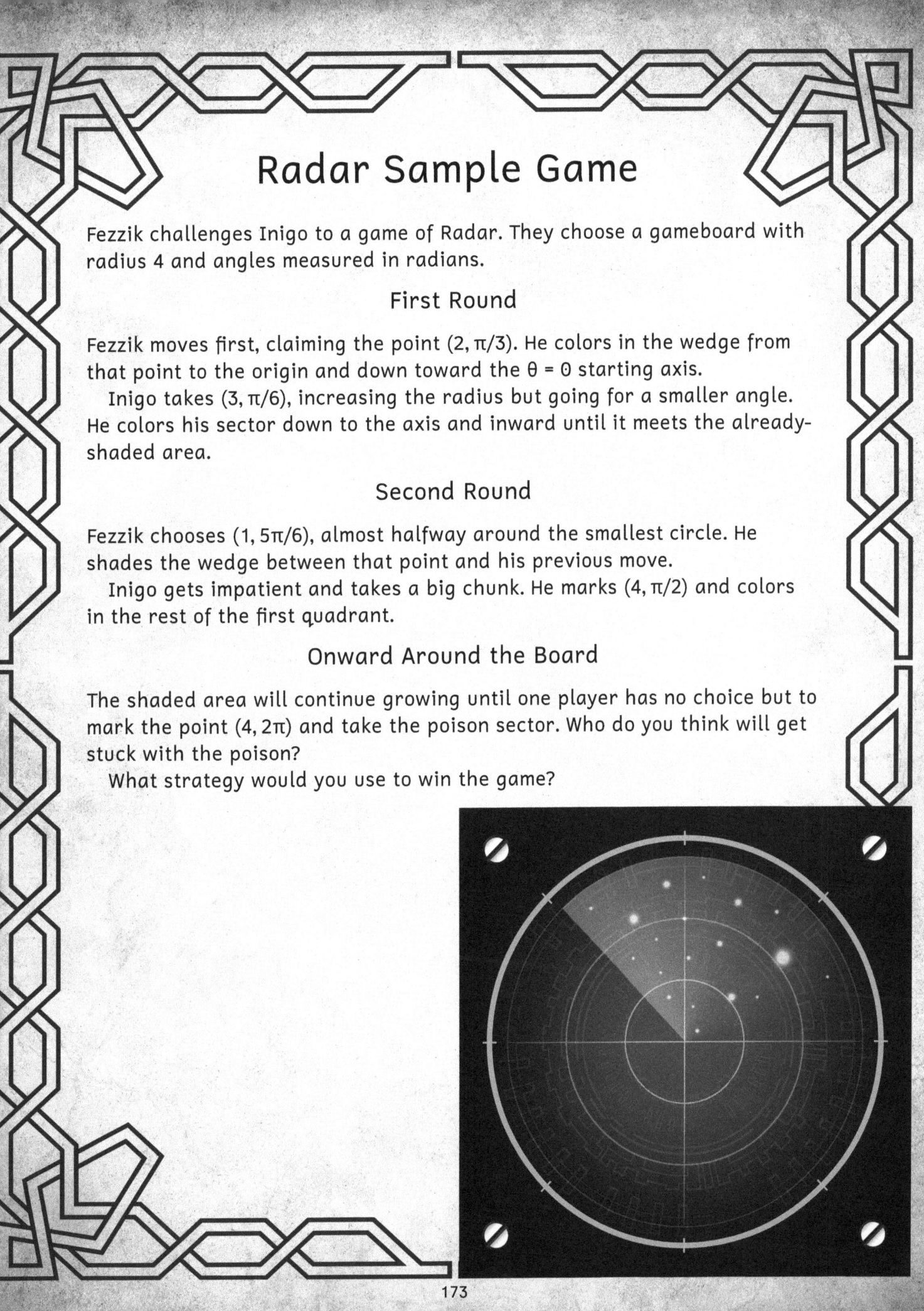

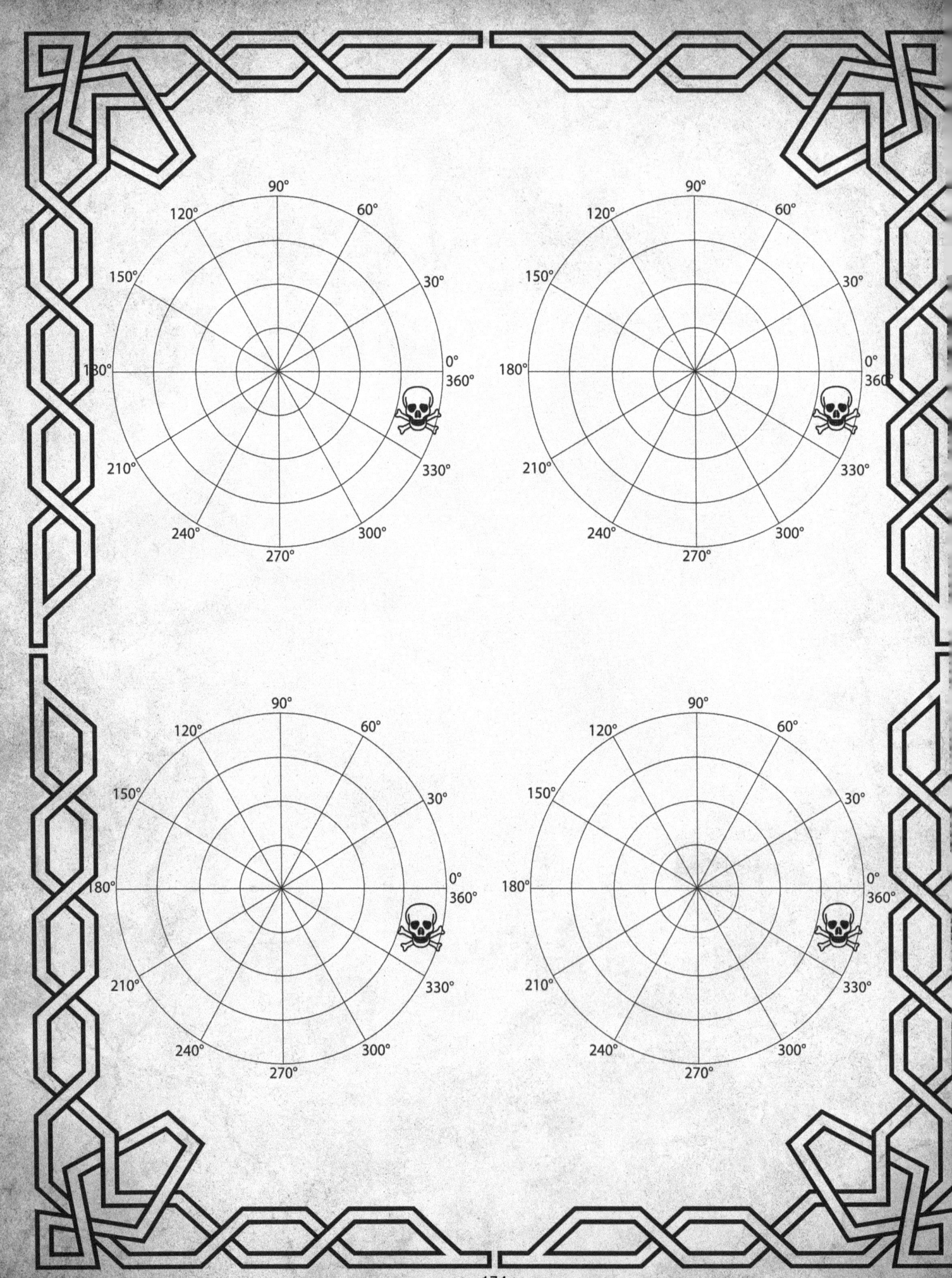

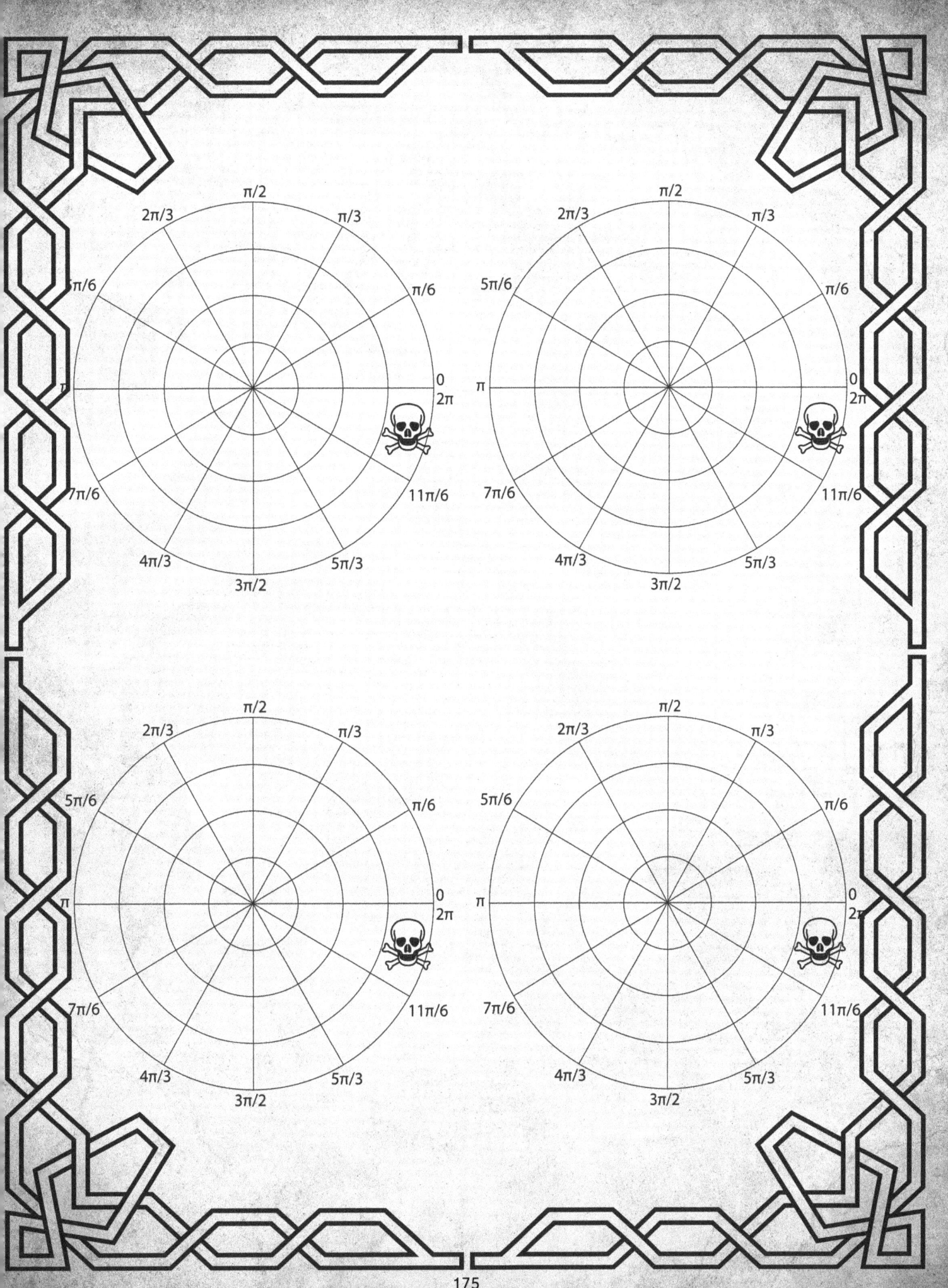

Notes & House Rules

Radian Race

MATH CONCEPTS: angle measures, polar coordinates.
PLAYERS: two to four.
EQUIPMENT: gameboard or polar (circular) graph paper, Small toys or other tokens.

Set-Up

Use a larger version of the Radar gameboard, in your choice of degrees or radians, but without the poison space.

Or make your own on polar graph paper. To create your own gameboard, mark circles of radius 1, 2, 3, etc. The number of circles must be equal to or greater than the number of people who want to play.

Around each circle, draw lines at convenient angles such as 30°, 60°, 90°, and so on. Or use radians: π/6, π/3, π/2, and so on.

How To Play

Each player chooses one circle and places a small game piece at θ = 0. On your turn, slide your piece around your own circle to the angle of the farthest player's position, and then move your choice of one or two angle sections further.

The first player to make a full trip around the circle, arriving back at the starting point, wins the game.

Variation

RADIAN NIM: Play as in the race above, but as a *misère* game. The first player to complete the circle loses.

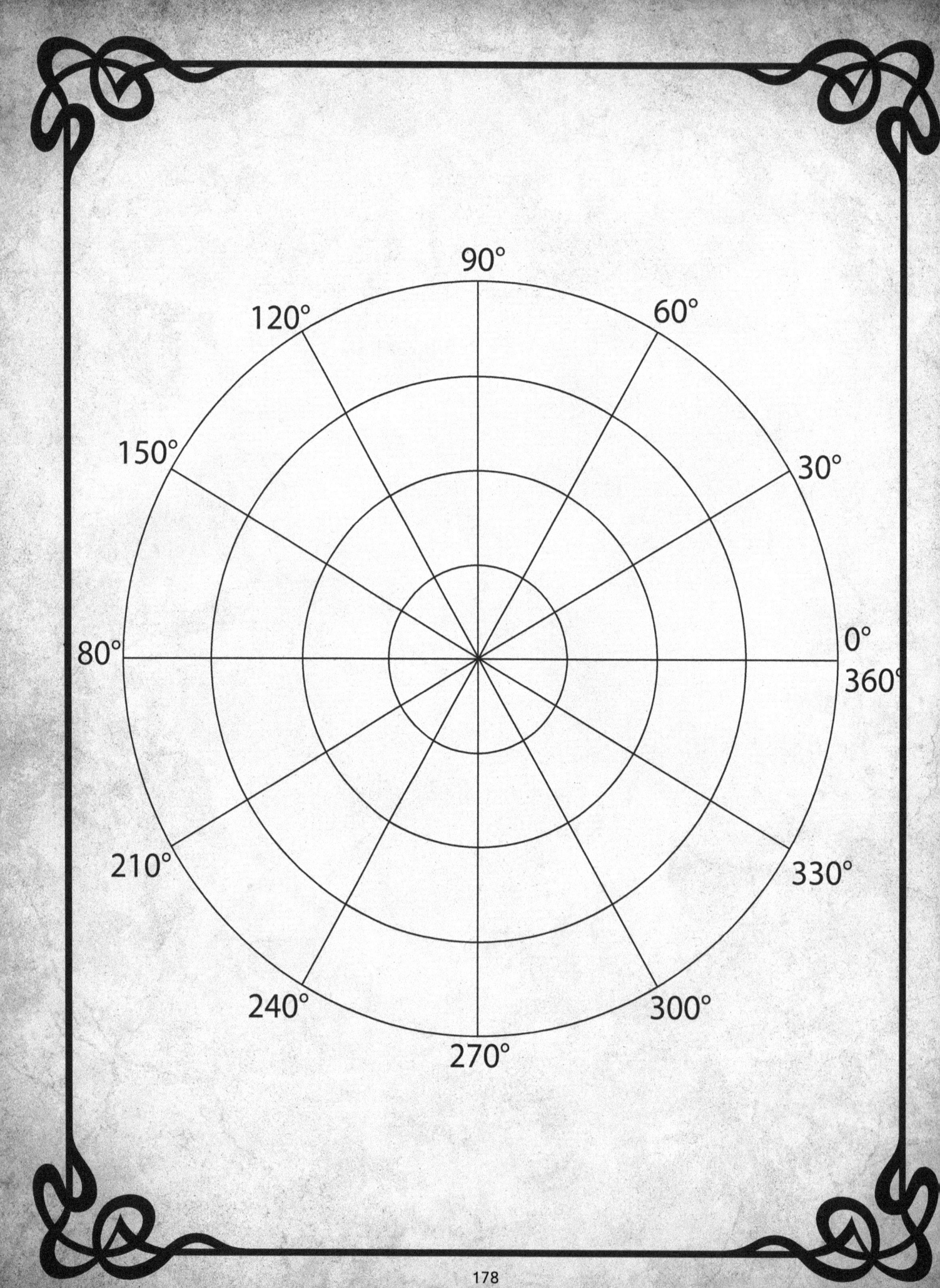

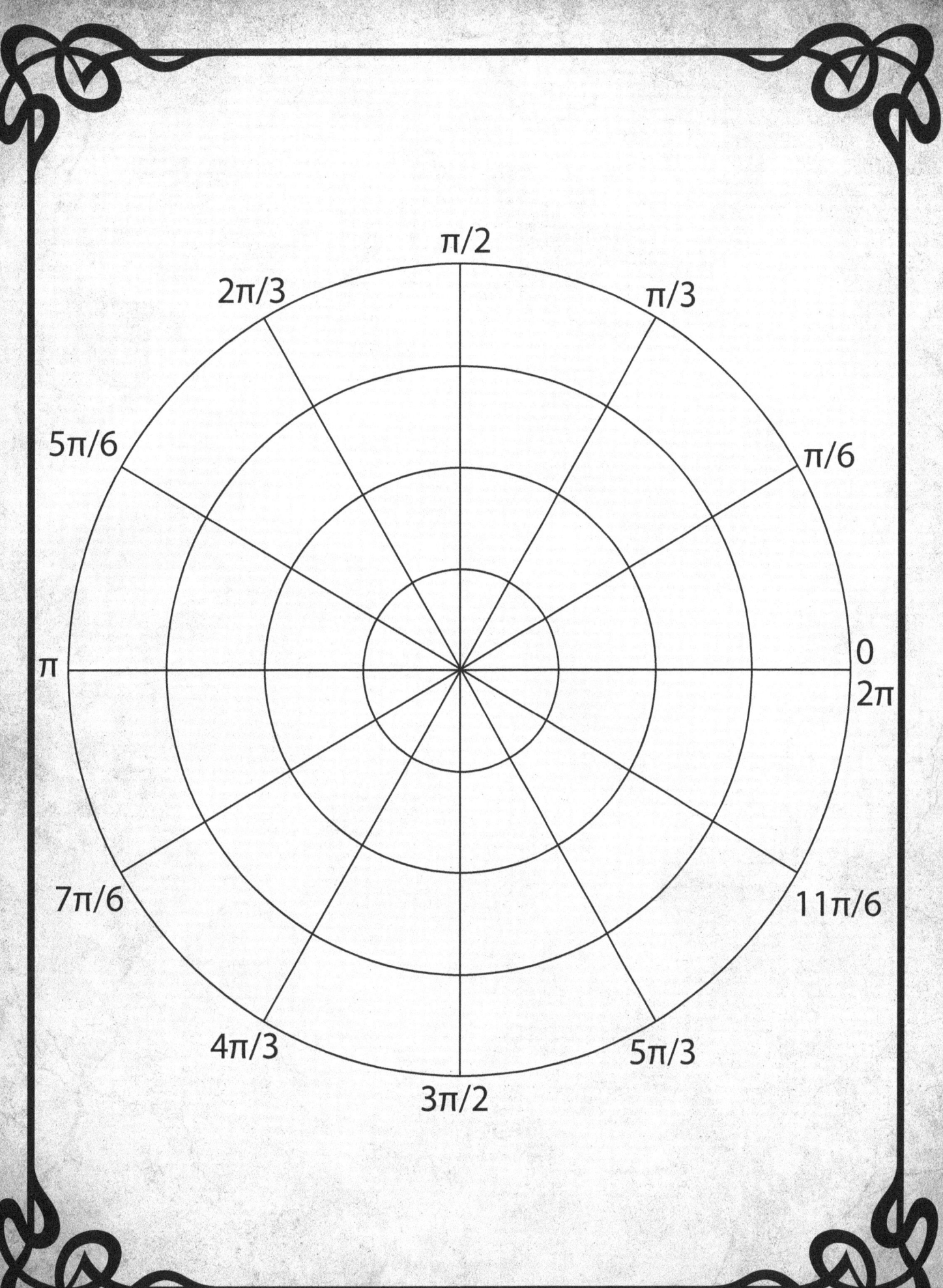

Notes & House Rules

Racetrack

Math Concepts: vectors, coordinate graphing, delta notation for changing coordinates.
Players: two or three.
Equipment: dotty or lined square grid paper, pencils or colored markers, ruler or other straightedge.

Set-Up

Make your own racetrack on grid paper by drawing an elliptical or irregular path. The track may have wide and narrow sections, but there must be at least two squares open between the walls at every point. Before the race starts, players should examine the track and decide whether any questionable points are on the road or out of bounds.

Draw two straight lines across the track to mark the start and finish lines for your race, or one line on a track that loops around. Also make a chart with a column for each player. Players choose a color or symbol (such as X and O) to represent their cars. Write the name and symbol at the top of each player's column.

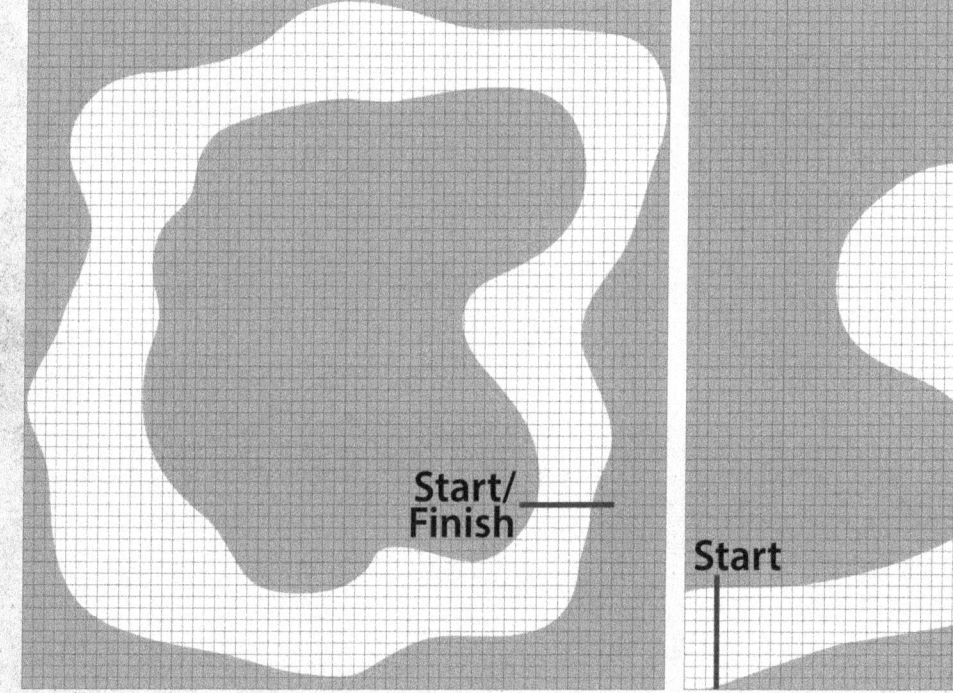
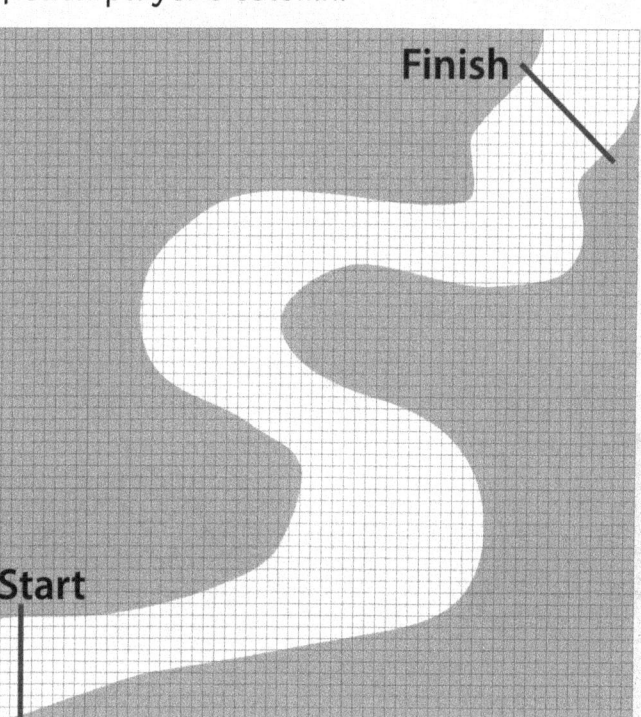

Two sample racetracks. You can race on a one-way track or around a loop.

How To Play

Each player marks a dot or grid intersection on or behind the starting line. Players move in the opposite order of how they chose their starting position—that is, whoever got first choice of position moves last.

On each turn, you travel along a vector $(\Delta x, \Delta y)$. This means you make some change in your horizontal position x, and some change in your vertical position y, measured relative to where you started that turn.

Racetrack Moves

For example, a vector of (−1, 2) means:

- Move −1 in the x direction, one square to the left.
- Move +2 in the y direction, two squares up on your grid paper.

Write the movement vector in your column on the gameboard. Then put a symbol at your new position and draw a straight line connecting it to your previous spot.

For your first turn, move one square in any direction. That is, you must choose the values for your initial movement vector from this list: −1, 1, or 0, mix or match. After that, you may change the Δx value, the Δy value, or both, by one square per turn. Or you can keep cruising at the same speed.

So after the (−1, 2) move described above, you might:

- Coast along with another (−1, 2) movement.
- Speed up to (−2, 3), two squares left and three squares up.
- Slow down to (0, 1), no horizontal movement and only one square up.
- Or any combination of these changes.

The lines that mark each player's path may cross, but a player may not move directly through another car's current position. Also, players must stay within the boundaries of the track for the entire length of their moves, or else they crash and lose the game.

The first person to travel around the track and cross the finish line wins the race. But any players who have not yet moved during that turn may complete their moves. If more than one person crosses the finish line in the same turn, then whoever goes the farthest past it is the winner.

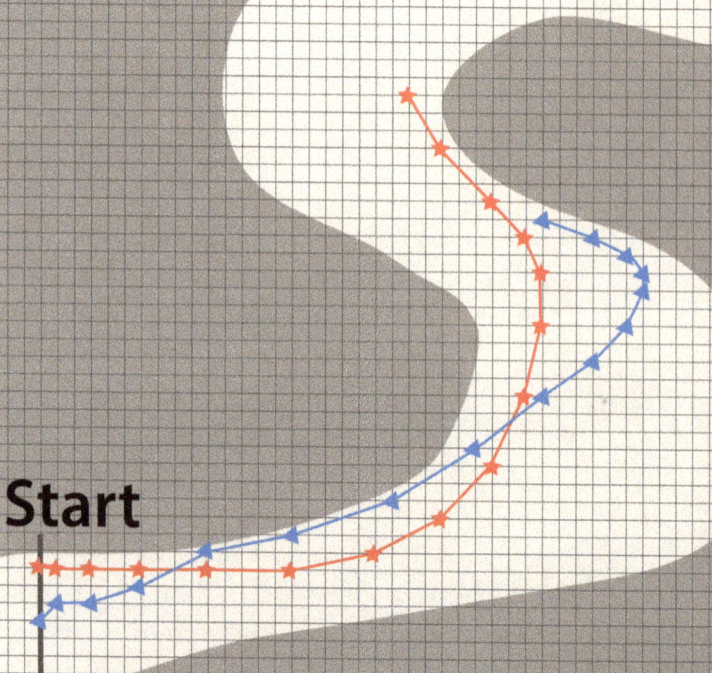

The beginning of a game of Racetrack, showing the movement vectors and each car's path.

The triangle player jumped ahead at first but went too fast and almost didn't make the turn.

★ Star	▲ Triangle
(1,0)	(1,1)
(2,0)	(2,0)
(3,0)	(3,1)
(4,0)	(4,2)
(5,0)	(5,1)
(5,1)	(6,2)
(4,2)	(5,3)
(3,3)	(4,3)
(2,4)	(3,2)
(1,4)	(2,2)
(0,3)	(1,2)
(−1,2)	(0,1)
(−2,2)	(−1,1)
(−3,3)	(−2,1)
(−2,3)	(−3,1)

Racetrack Hazards (optional)

Before the race starts, players may lightly color sections of the track to mark them as hazards. For instance, you might try...

OIL SPILLS: Shade light gray with pencil. Cars traveling through an oil spill cannot get traction to accelerate. If you pass through this area, your movement vector must repeat your previous turn.

LOOSE GRAVEL: Fill the area with small dots. Gravel slows you down and limits your control. When you drive on gravel, both coordinates of your movement vector must be half the value of your previous turn. With an odd coordinate, you may round the half-value up or down. If you come to a complete stop, you may move one square on your next turn, as at the start of the race.

TURBO BOOST: Fill the area with small arrows all pointing the same direction. Within this area, your movement vector increases (or decreases) by two squares, but only in the direction of the arrows. Horizontal arrows change the Δx value, vertical arrows the Δy value, and diagonal arrows change both. A coordinate not affected by the turbo arrows follows the normal rule.

Other Variations

HOUSE RULE: Do you hate to lose the game because of a crash? Choose a rule that gives players a chance to recover from miscalculations.

For example, players who drive off the track can move one square per turn until they return to the racetrack.

Or the player moves one square per turn, and the car must reenter the track behind the place where it went off.

Or players may return magically to any point behind where they left the track. They lose their next turn—a $(0, 0)$ movement vector. After that, they accelerate as at the beginning of the game.

About the Math

A *vector* is a force or movement that has both magnitude and direction. When we write polar coordinates, we are writing vectors. The coordinates (r, θ) represent that point's (*magnitude*, *direction*) from the origin point $(0, 0)$.

We can also use vectors on an x-y Cartesian coordinate grid. In that case, we don't write the magnitude and direction precisely, but instead show the relative change in position.

In the Racetrack game, we represent a movement vector (also called a *translation*) by writing an ordered pair $(\Delta x, \Delta y)$. Read this as "delta-x, delta-y." The Greek capital delta (Δ) means *the change in*. So Δx is the change in horizontal x position. And Δy is the change in vertical y position.

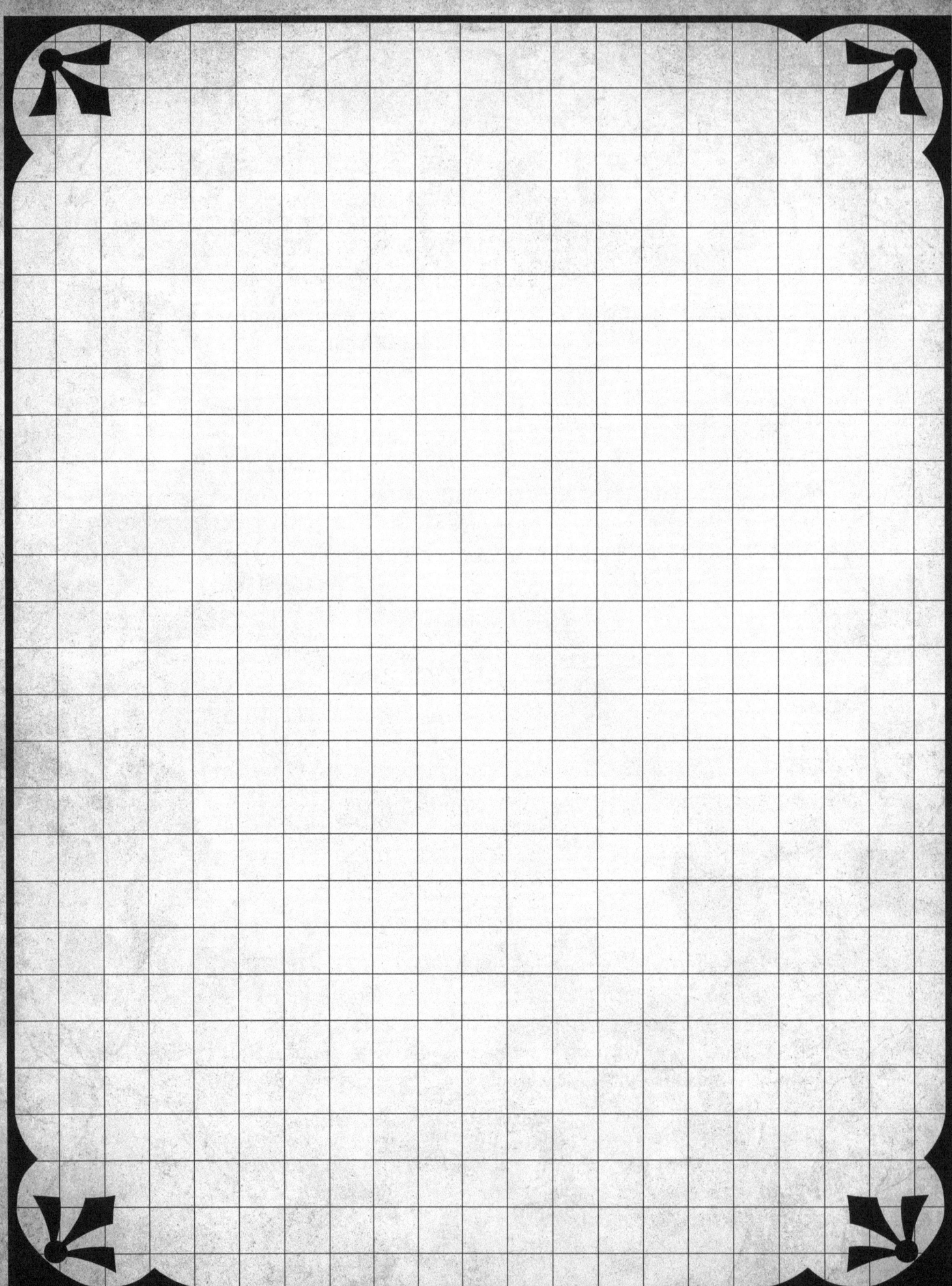

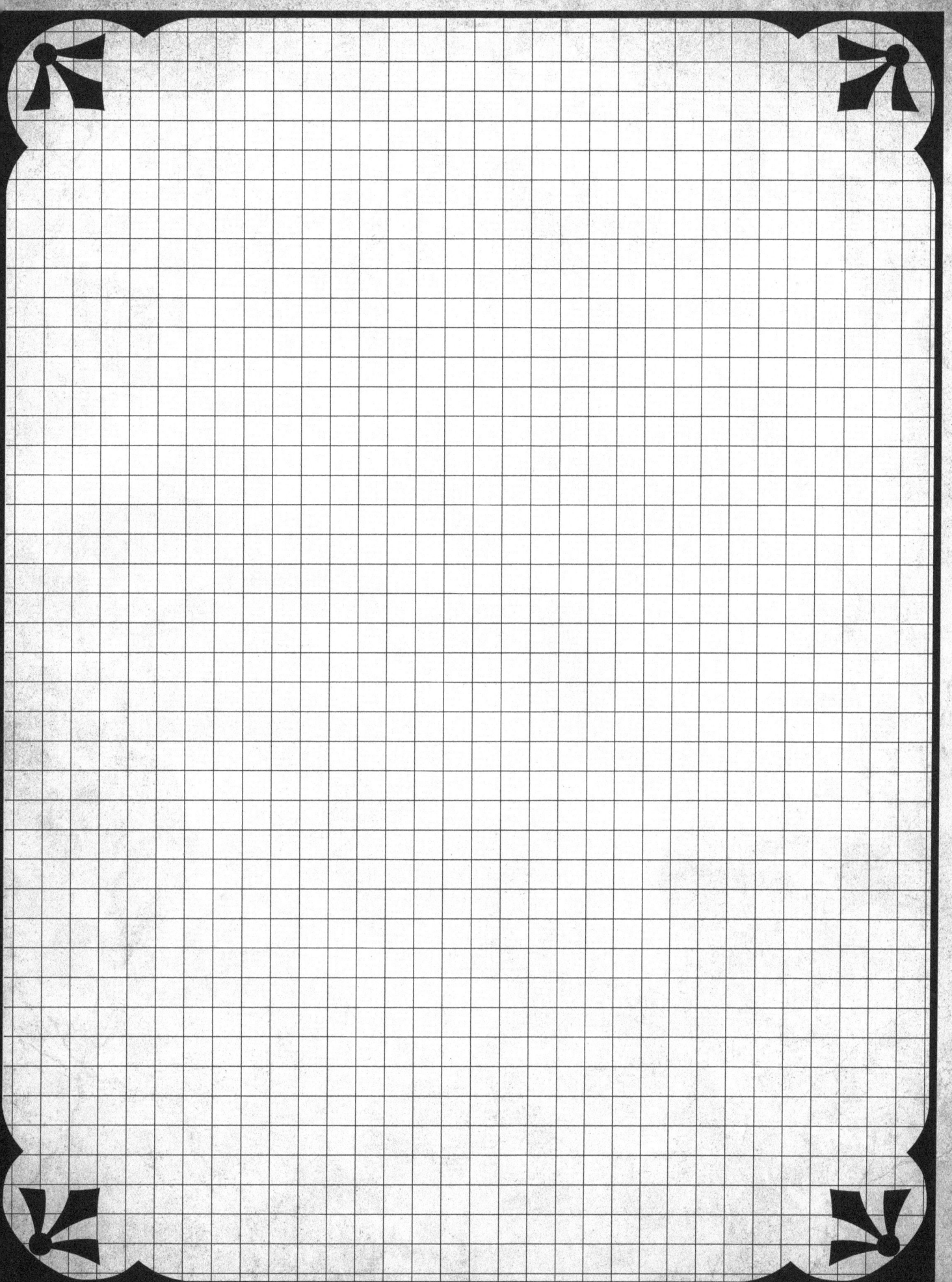

For More Information

Hidden Hexagons

Teacher coach Terry Kawas modified the traditional Battleship game to create a quadrilateral guessing game called Shape Capture. Teacher and author Christopher Danielson shared a wonderful geometry investigation called Hierarchy of Hexagons. If ideas can have children, this game of mine is their offspring.

- ♦ christopherdanielson.wordpress.com/?s=hexagons

Linear War

This is yet another game from the inventive mind of John Golden, the Math Hombre. He suggests using it throughout a unit on linear graphing, having students make different game cards as you introduce each vocabulary term. At the end of the unit of study, they can choose their eleven-card decks and play the game for review.

- ♦ mathhombre.blogspot.com/2011/07/linear-war.html

Radar

Radar is my circular version of the two-player strategy game Chomp, which inventor David Gale described as a curious way to eat a chocolate bar. Author Martin Gardner popularized the game.

Gale, David. "A Curious Nim-type Game," *The American Mathematical Monthly*, vol. 81, issue 8, 1974.

Gardner, Martin. "Sim, Chomp and Race Track: New Games for the Intellect (and Not for Lady Luck)," Mathematical Games column in *Scientific American* magazine, January 1973.

Racetrack

Racetrack is also called Vector Race or Graph Paper Race. In 1973, Martin Gardner wrote about the game for his "Mathematical Games" column in *Scientific American*. Michael Serra explained how to use vector notation for player moves, Dan Meyer blogged about it, and I saw it on Meyer's dy/dan blog.

- ♦ blog.mrmeyer.com/2008/asilomar-5-michael-serra

Number Neighborhoods

TWO MATH GAMES OF LOGICAL DEDUCTION

Notes & House Rules

Which Number Where?

MATH CONCEPTS: arithmetic, number properties, deductive thinking.
PLAYERS: two players or two teams.
EQUIPMENT: gameboard (optional), pencils and paper, or whiteboards and markers.

In this game, each player (or team) controls a small group of three houses. The numbers one through eight live in these houses, and your goal is to discover where the other player's numbers are hiding.

Set-Up

Each player/team draws a gameboard with two rows of three houses each. Draw a horizontal line to make two floors in each house: Top and Bottom.

Label one row of houses Mine and the other one Theirs.

Label the houses in each row Left, Middle, and Right.

Divide the floors into rooms: Left and Right have three rooms on each floor, while the Middle house has two rooms on each.

How To Play

Write the numbers 1–8, each in a separate room in your own row of houses. Some rooms will be empty. Cover your neighborhood with a sticky note, or fold the paper to keep the numbers hidden.

On your turn, ask the other player any Yes or No question. For example:

- Is 1 in a top room?
- Are there any empty rooms?
- Are there any rooms with two even numbers?

The other player answers the question truthfully with either a Yes or a No (and nothing else). You can write notes to help you figure out what you know from the information you have so far.

If the answer to your question is *yes,* then you get another turn. If the answer is *no,* that's the end of your turn.

Endgame

On your turn, if you think you know in which house and on which floor all of the other player's numbers live, then ask one at a time if each number is in the location you think. If the answer is *yes* to all eight questions, then you win! (If not, then at the first *no,* it's the other player's turn and the game continues.)

Which Number Where?

My Neighborhood: Their Neighborhood:

LEFT MIDDLE RIGHT   LEFT MIDDLE RIGHT

TOP
BOTTOM

Clues:

Which Number Where?

My Neighborhood:

LEFT MIDDLE RIGHT

Their Neighborhood:

LEFT MIDDLE RIGHT

TOP
BOTTOM

Clues:

Notes & House Rules

Number Neighborhoods

MATH CONCEPTS: sets, open intervals, deductive thinking.
PLAYERS: two players or two teams.
EQUIPMENT: gameboards, scratch paper, pencil or pens.

In this game, each player (or team) controls three number neighborhoods, which are intervals along the number line between zero and ten. Your goal is to discover where the other player's numbers are hiding.

Set-Up

Players may create their own gameboards.

Choose six different numbers between zero and ten, exclusive. Each number may have at most one decimal place.

For example, you might choose 0.8, 3.2, 5.6, 5.9, 6.0, 8.7.

Write the numbers in increasing order to name the union of three sets. For example, the numbers above would look like this:

$$(0.8, 3.2) \cup (5.6, 5.9) \cup (6.0, 8.7)$$

This parenthetical notation represents a large set on the number line made of three separate *open intervals*—sets that don't include their end points. (If we wanted the intervals to be *closed*, inclusive of their endpoints, we'd use square brackets.)

Color the intervals on your number line.

The example intervals would look like this:

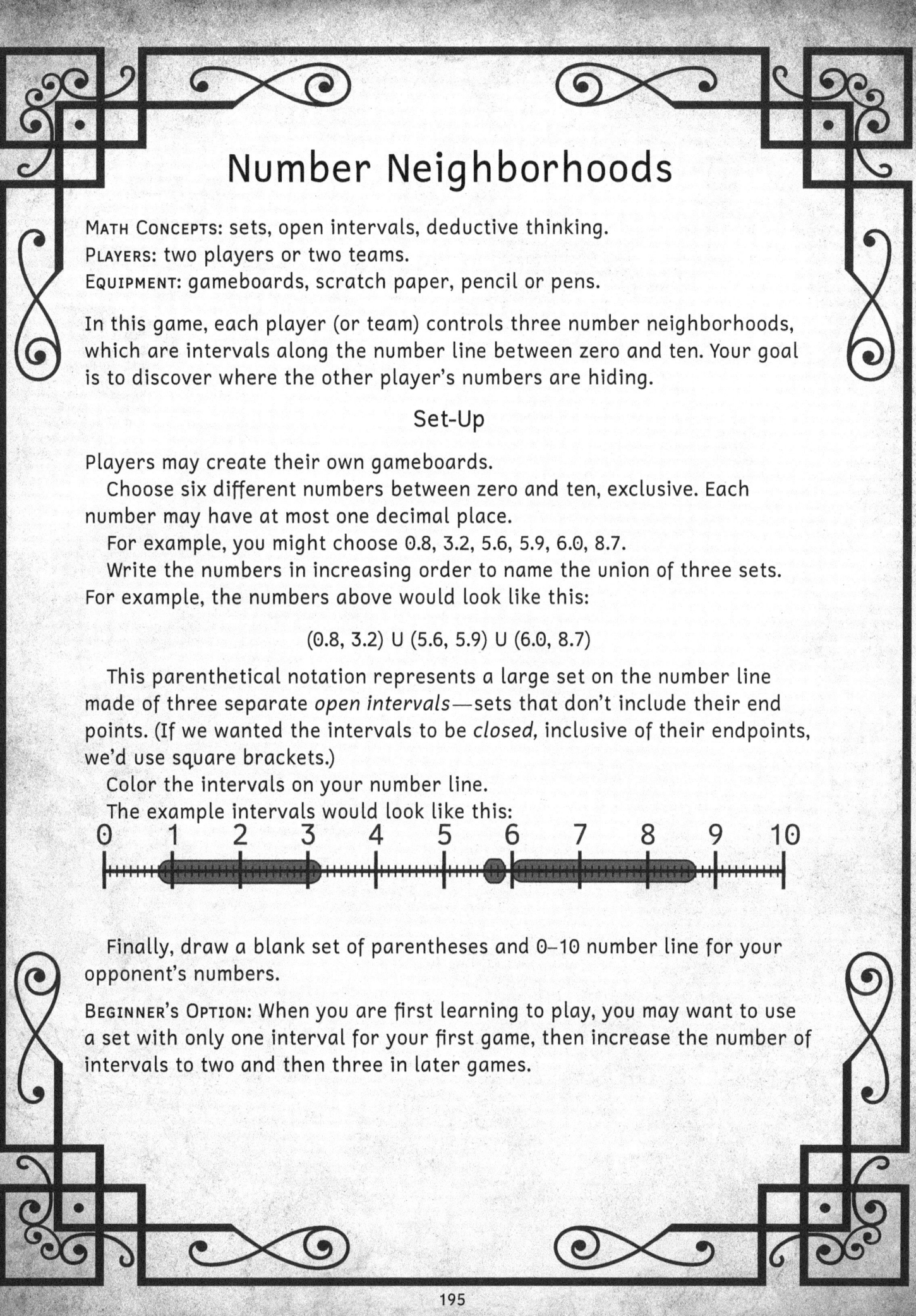

Finally, draw a blank set of parentheses and 0–10 number line for your opponent's numbers.

BEGINNER'S OPTION: When you are first learning to play, you may want to use a set with only one interval for your first game, then increase the number of intervals to two and then three in later games.

How To Play

On your turn, ask for up to five sets of numbers in the format "all the numbers less than D away from C." These intervals must all have the same center C, and the distance D must be smaller each time.

For example, you could name the following sets on one turn:

- All the numbers less than 2 away from 6.5.
- All the numbers less than 1 away from 6.5.
- All the numbers less than 0.5 away from 6.5.
- All the numbers less than 0.2 away from 6.5.
- All the numbers less than 0.1 away from 6.5.

You don't have to use all five questions in a turn. But if you want to ask about a different center or a wider distance, you have to wait until your next turn.

After you ask for a set, the other player responds:

- All of those numbers are inside my set. (Or say "inside" for short.)
- All of those numbers are outside my set.
- Some of those numbers are inside my set, and some are outside. (Or say "both.")

All sets in this game are open—they don't include the endpoints of any of the intervals. If the whole interval except the endpoints is inside one part of their set then the answer is "inside," and if the whole interval except the endpoints is outside all three parts of their set, then the answer is "outside."

Write notes to keep a record of the intervals you have guessed so far and their response clues.

Endgame

Once during the game, on your own turn and instead of asking about interval sets, you can say you are ready to guess your opponent's numbers. Then you say the six endpoints of their sets.

If you are right for all six numbers, you win! If you are wrong for even one number, you lose and the other player/team wins. Either way the game is over.

A Sample Game

Elizabeth played a game of Number Neighborhoods with her sister Jane. Elizabeth claimed the sets we mentioned earlier as examples:

(0.8, 3.2) U (5.6, 5.9) U (6.0, 8.7)

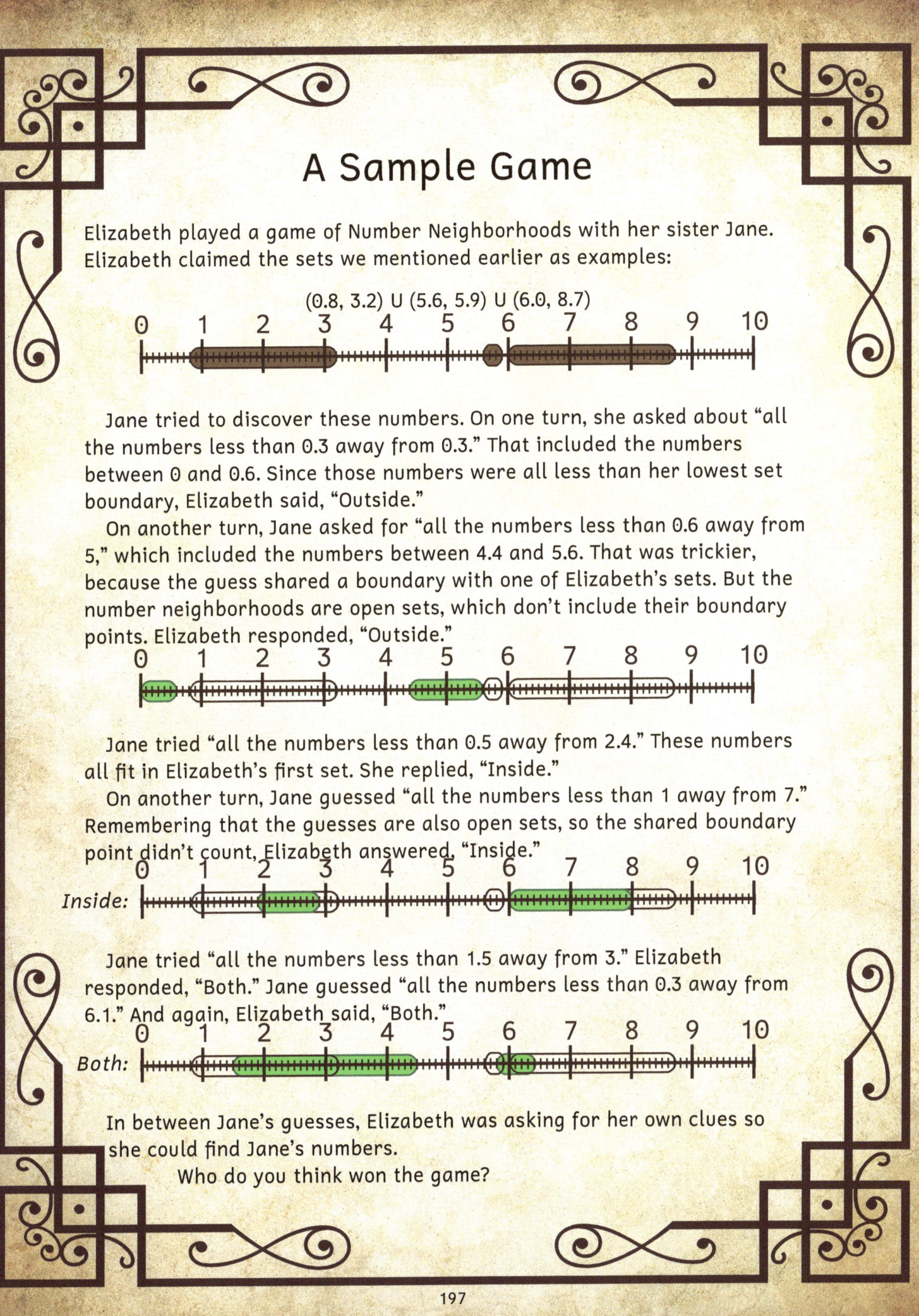

Jane tried to discover these numbers. On one turn, she asked about "all the numbers less than 0.3 away from 0.3." That included the numbers between 0 and 0.6. Since those numbers were all less than her lowest set boundary, Elizabeth said, "Outside."

On another turn, Jane asked for "all the numbers less than 0.6 away from 5," which included the numbers between 4.4 and 5.6. That was trickier, because the guess shared a boundary with one of Elizabeth's sets. But the number neighborhoods are open sets, which don't include their boundary points. Elizabeth responded, "Outside."

Jane tried "all the numbers less than 0.5 away from 2.4." These numbers all fit in Elizabeth's first set. She replied, "Inside."

On another turn, Jane guessed "all the numbers less than 1 away from 7." Remembering that the guesses are also open sets, so the shared boundary point didn't count, Elizabeth answered, "Inside."

Jane tried "all the numbers less than 1.5 away from 3." Elizabeth responded, "Both." Jane guessed "all the numbers less than 0.3 away from 6.1." And again, Elizabeth said, "Both."

In between Jane's guesses, Elizabeth was asking for her own clues so she could find Jane's numbers.

Who do you think won the game?

Number Neighborhoods

My Sets:

(_____ , _____) ∪ (_____ , _____) ∪ (_____ , _____)

0 1 2 3 4 5 6 7 8 9 10

Their Sets:

(_____ , _____) ∪ (_____ , _____) ∪ (_____ , _____)

Clues:

0 1 2 3 4 5 6 7 8 9 10

Number Neighborhoods

My Sets:

(_____ , _____) ∪ (_____ , _____) ∪ (_____ , _____)

0 1 2 3 4 5 6 7 8 9 10

Their Sets:

(_____ , _____) ∪ (_____ , _____) ∪ (_____ , _____)

Clues:

0 1 2 3 4 5 6 7 8 9 10

For More Information

The games Which Number Where? and Number Neighborhoods were created by David Butler, a lecturer at the Maths Learning Centre, University of Adelaide. You can read more about each game (and download printable battleship-style gameboards) at his blog:

- ♦ Which Number Where?
  adelaide.edu.au/mathslearning/news/list/2020/08/18/which-number-where

- ♦ Number Neighborhoods
  adelaide.edu.au/mathslearning/news/list/2020/09/05/number-neighbourhoods

Butler runs the weekly puzzle session called One Hundred Factorial, named after the classic puzzle that launched the event:

The number 100! (pronounced "one hundred factorial") is the number you get when you multiply all the whole numbers from 1 to 100.
That is, 100! = 1 × 2 × 3 × ... × 99 × 100.
When this number is calculated and written out in full, how many zeros are on the end?

Butler shares his games at these puzzle sessions, which always lead to great conversations about what people are thinking or wondering: How many different ways can we approach a question? What happens if we tweak the parameters of a puzzle or game? Where else might this path of reasoning lead?

♦ ♦ ♦

"I love the games even more when we play in teams. I adore listening to the discussion between the players in a team about what question to ask next, and about what information they have from the questions they have asked so far.

"It's always fascinating to hear people's thinking, and a game of pure logic is such an easy way to get that thinking on display.

"I hope you enjoy playing (and listening to others playing) as much as I do."

—David Butler, "Which Number Where?"

Special Thanks

This book came to production with the help of many wonderful people who backed the project on Kickstarter. Your generous support and encouragement keeps me going!

I'm especially grateful to:

- Alex Joel
- Amanda Bartonek
- Amy Bloyer
- Amy Chen
- Amy Diamond Weiny
- Annabelle, Will, Anya, John, Jacob, and Michal Lee
- Bekka Christophel
- Beth Manz
- Brittney Sawka
- Callister Family
- Calvin, Malcolm, and Spencer Faull
- Cam C
- Canner Family
- Carles Bona
- Carlos Gonzalez
- Christopher K. Onken
- Clair J
- Costa
- Donovan Hayes
- Douglas & Molly Naaden
- Elio
- Elizabeth Cousin
- Feliciano Family
- Finley Clover Adcock
- Geoffrey M Allen
- Gerben Wieringa
- Gustin Family
- Halstead Family
- Jefe
- Jen Campbell
- Jill Marie Hubbard
- Jo Craig
- Jo Oehrlein
- JQ
- Julie V
- June Turner
- Kate Silgals
- Katie & Caillie
- Kelli Poll
- Kevin Glenn
- Kirk's Tutoring
- Krystal Bohannan
- Kyle & Stephanie Stephens
- Lauren Osborne
- Lewis Burridge
- Lisa Ferland
- Louis Luangkesorn
- Loxie
- Lynne Menechella
- Margie White
- Matthew Bull
- Michelle Scharfe
- Milo Kotowski
- Mindie and Drew Simmons
- Mrs. D
- Myrhat Eliot
- Nacho Ruiz Cía
- Nancy Fox
- Nimish Shah
- Pattie Perry
- Rachel Donnell
- Raman Ladutska
- Riley Family
- Rineer Family
- RL
- Robbie Long
- Robin Wienke
- Saleh M. Abdullah
- Santos Family
- Shawnda Lindsey
- Shinya Yoshida
- Steven Toon
- Sykora Family
- the Burkitos
- The Karimeddinis
- Tracy Popey
- Tully Family
- Vic Beaumont
- Walter Thomas
- Wendy Maynard

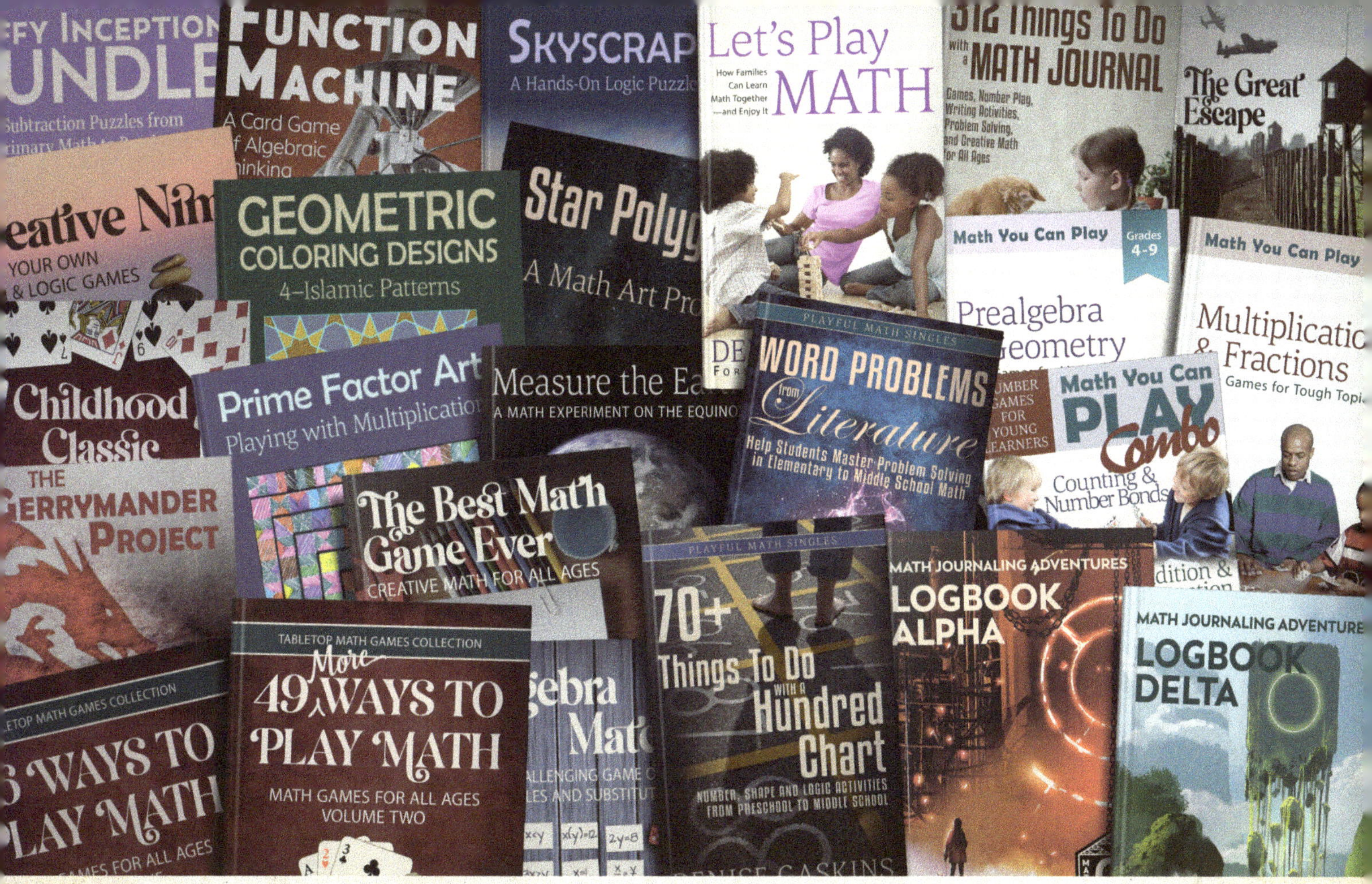

Playful Math Books by Denise Gaskins

- If you're a parent trying to help your child learn math...
- Or a teacher looking for creative ideas for your classroom...
- Or a homeschooling parent hoping to enrich your student's understanding...

Then you'll love how Denise's books and activity guides lead you and your children to explore mathematics at a deeper level, building a strong foundation to support future learning.

Playful math enriches any curriculum. It doesn't matter whether your students are homeschooled or in a classroom, distance learning or in person.

Because everyone can enjoy the experience of playing around with math!

Tabletop Academy Press publishes playful math books and cool mathy merchandise for parents who want to help their children build the understanding and skills they need to succeed in school and beyond.

Homeschoolers, afterschoolers, unschoolers, and even classroom teachers appreciate our flexible approach that can work alongside any math program.

Visit us today:
TabletopAcademyPress.com

Or browse Denise's blog:
DeniseGaskins.com

www.ingramcontent.com/pod-product-compliance
Lightning Source LLC
Chambersburg PA
CBHW051330110526
44590CB00032B/4473